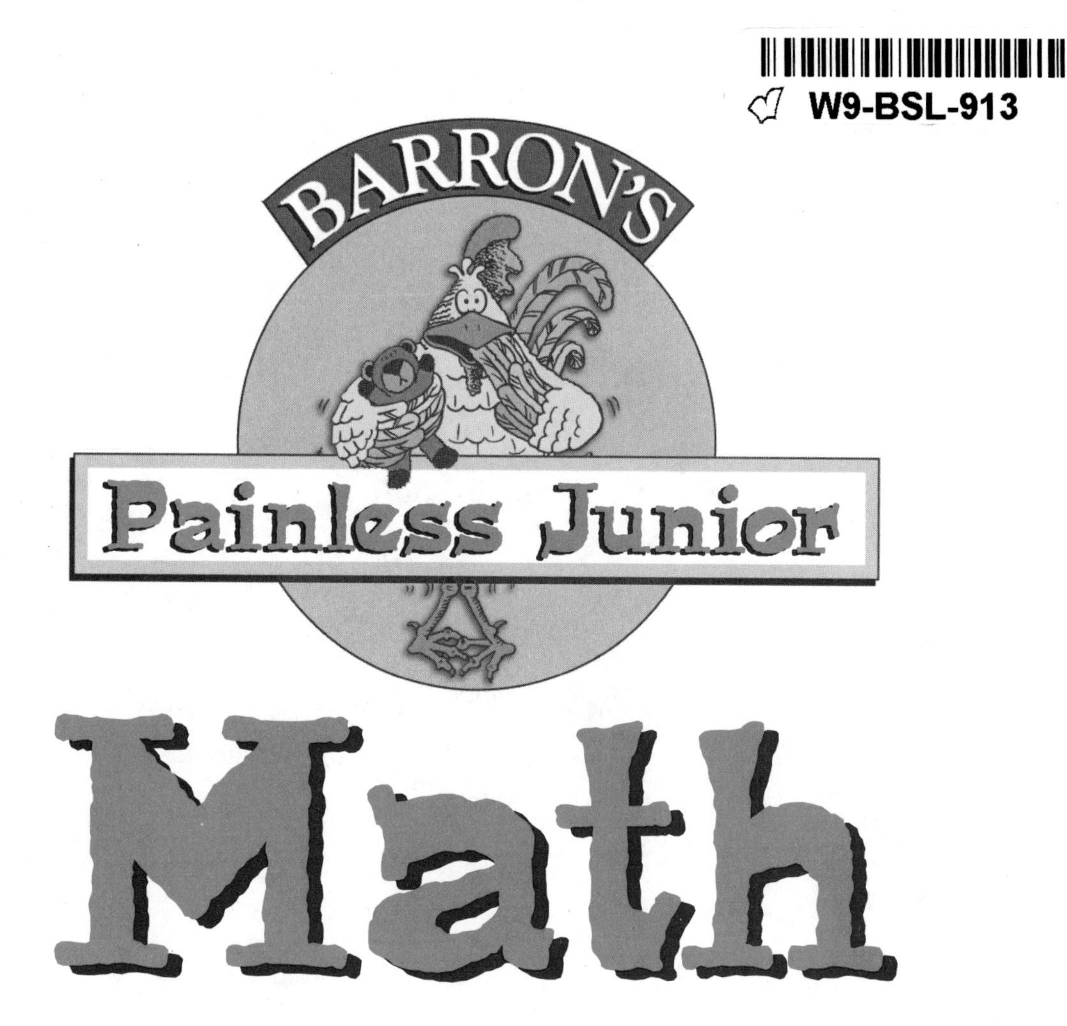
BARRON'S
Painless Junior
Math

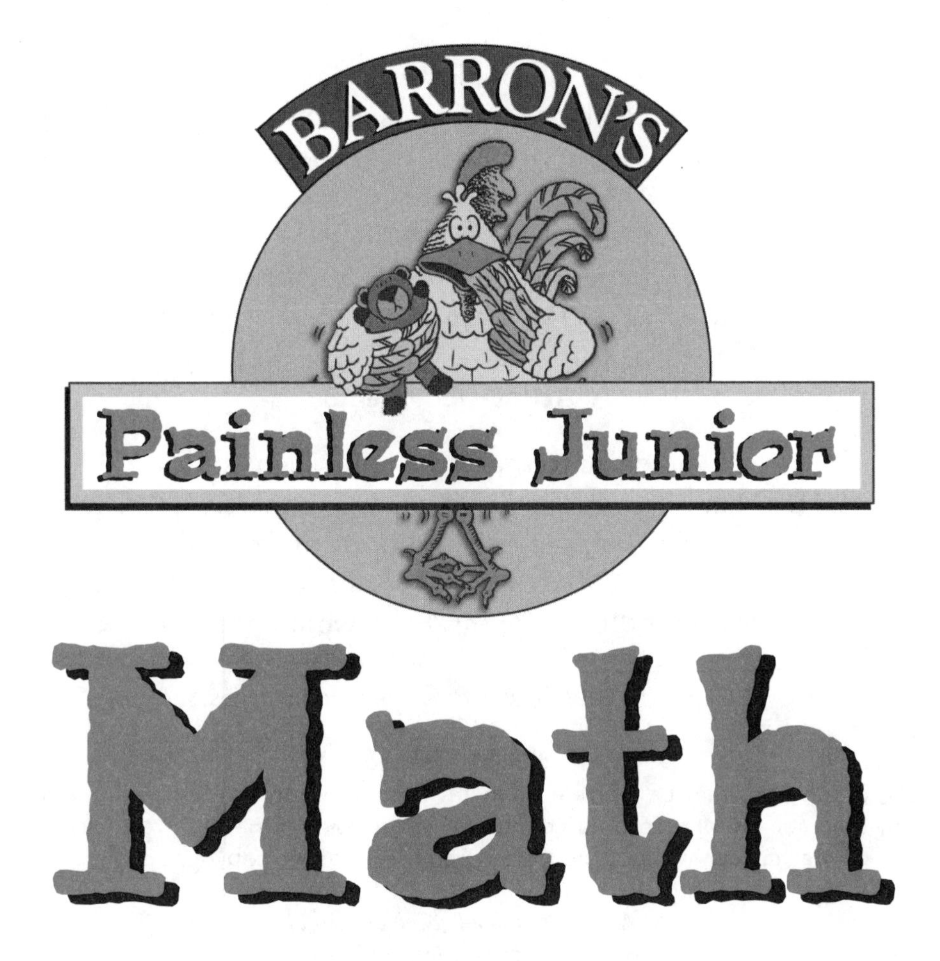

Margery Masters, M.Ed.

Dedication

This book is dedicated to my parents, Ona and Arnold Masters,
whose love and support has always been there for me.

About the Author

Margery Masters is the author of
Barron's *Let's Prepare for the Grade 4 Math Test*.

Acknowledgments

I would like to thank the following people:
Ed Brennan, Brenda Clarke, Diane Deger, Loren Finkelstein,
Anne Haring, Keith and Cody Krzyzewski,
Ona Masters, and Susan Yoder.
Without their help, this book would not exist.

All inquiries should be addressed to:
Barron's Educational Series, Inc.
250 Wireless Boulevard
Hauppauge, New York 11788
www.barronseduc.com

Library of Congress Catalog Card No. 2006042846

ISBN-13: 978-0-7641-3450-0
ISBN-10: 0-7641-3450-7

Library of Congress Cataloging-in-Publication Data
Masters, Margery.
Painless junior: math / Margery Masters.
p. cm.
Includes index.
ISBN-13: 978-0-7641-3450-0 (alk. paper)
ISBN-10: 0-7641-3450-7 (alk. paper)
1. Mathematics—Study and teaching (Elementary)
2. Mathematics—Examinations, questions, etc.
I. Title.
QA135.6.M3736 2006
510—dc22 2006042846

Printed in the United States of America
9 8 7 6 5 4 3 2 1

Contents

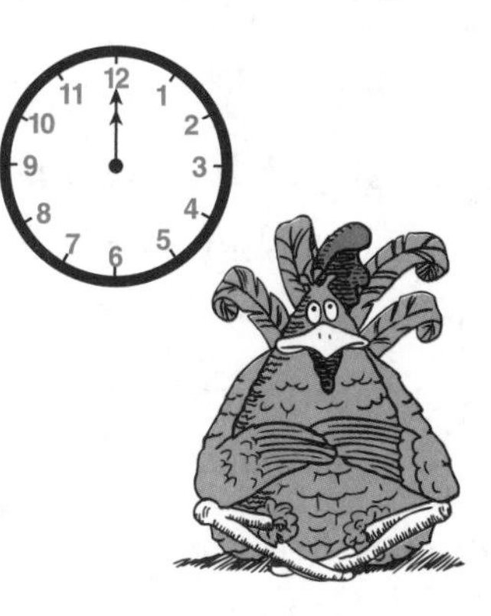

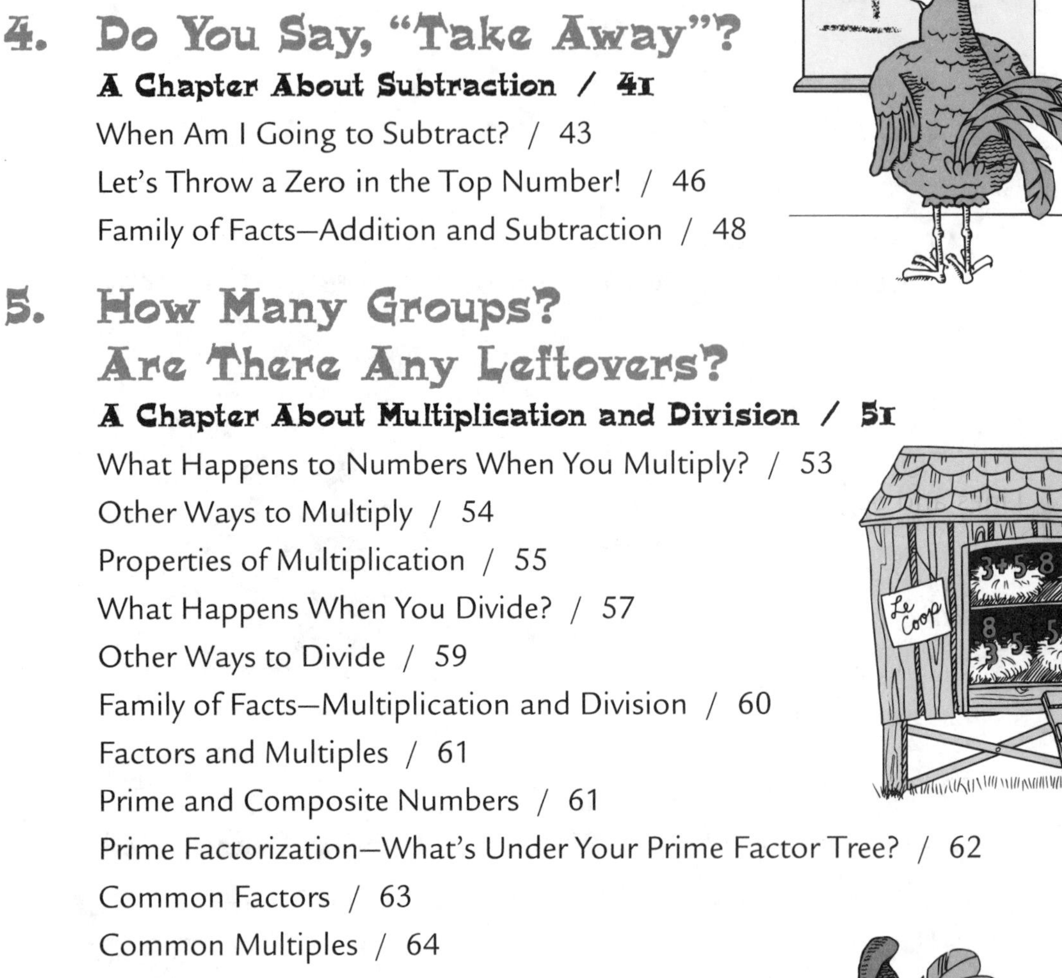
Le Coop

Icon Key

What You'll Find...
in the chapter.

Careful!
Watch our for possible problems.

Think About It!
A question to test your understanding.

Let's Try It!
A quiz to check what you just learned.

Important!
Pay attention to these facts you must know.

Chapter 1
What Time Is It?
A Chapter About Telling and
Understanding Time

Terms and Definitions

Analog clock: A device for measuring time, using hands that move in a circle showing hours, minutes, and sometimes seconds.

Digital clock: A device for measuring time that shows hours and minutes in digit form.

Minute hand: The minute hand is the longer hand that points to the minute on an analog clock.

Hour hand: The hour hand is the shorter hand that indicates the hour on an analog clock.

Elapsed time: The amount of time that passes from the beginning of an activity to the end of it.

A.M.: The time between midnight and noon.

P.M.: The time between noon and midnight.

IT'S ABOUT TIME

Have you ever wondered when people started telling time? How often do you look at the clock during the day? I'll bet you look at the clock more often in school than when you are home or playing with your friends. Have you ever wondered why that is true? Perhaps we are more aware of the time when our teachers control it than when we get to do what we want at home or with friends.

Our friend Pethagerus is here to help us.

He is learning about time, too!

Here are some clocks you might see every day.

This is a digital clock.

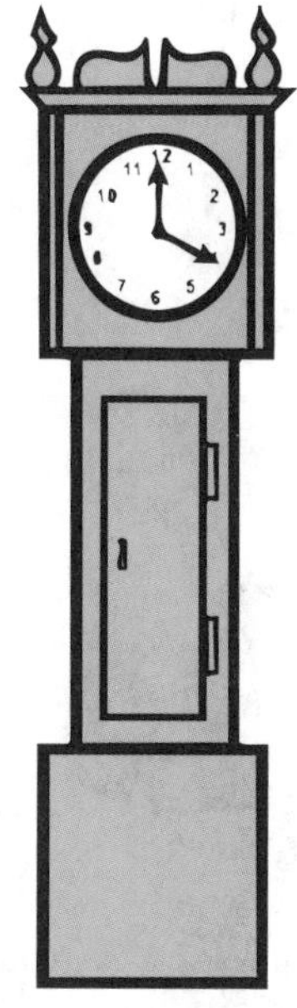

This is an analog clock.

TIME TO THE HOUR

We'll start at the beginning.
This clock is showing 4 o'clock.
The longer hand, called the **minute hand**, is pointing to the 12 and the shorter hand, called the **hour hand**, is pointing to the 4. Every time the minute hand points to 12 a new hour begins. The hour hand tells you what hour that is!

Let's Try It! Set #1

What time is shown by each of these clocks?

Think About It!

What time is shown when both hands are pointing straight up?

Draw your own hands on the clock to help you answer this question.

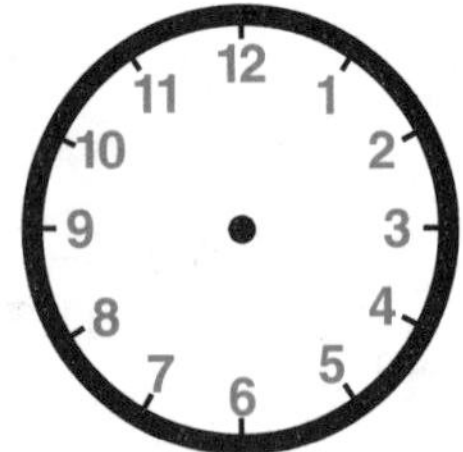

Answers are on page 171.

TIME TO THE HALF HOUR

Now that you know how to tell time on the hour, let's look at time to the half hour. One half hour is 30 minutes. The minute hand travels halfway around the clock to show this time. Look at the clock.

When you add the hour hand it looks like this.

The time shown is 3:30.

What do you notice about the hour hand?

Careful!

Remember that as the time changes each hour, the hour hand travels from one hour to the next.

So at 3:30 the minute hand is pointing to the 6 and the hour hand is halfway between the 3 and the 4.

Let's Try It! Set #2

What time is shown by each of these clocks?

1. 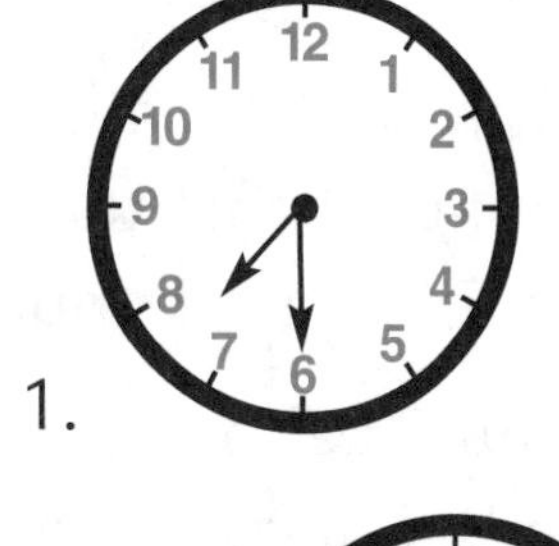

2. 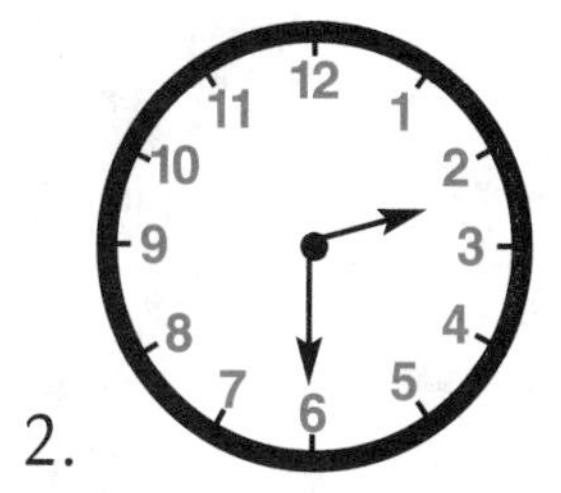

3. 

4. 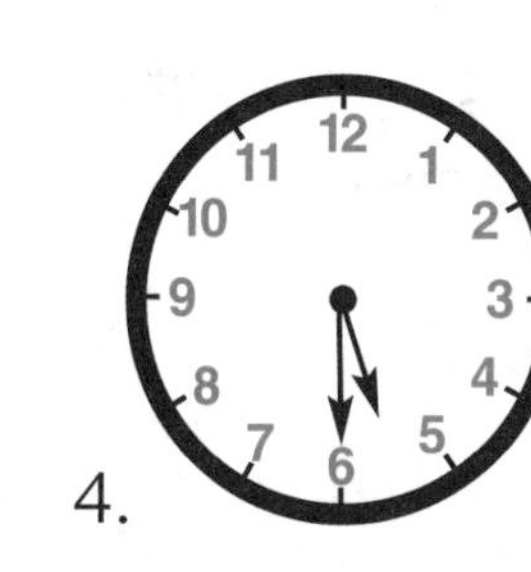

5. 

Think About It!

Where are the hands on a clock that shows 6:30?

Draw the hands on the clock to show 6:30.

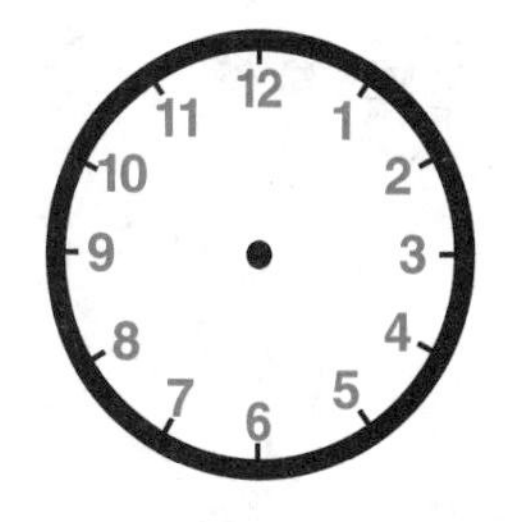

Answers are on page 171.

TIME TO THE MINUTE

OK! So now that we have learned that, what happens for the rest of each hour? Look at the clock.

It isn't 8 o'clock but it isn't 8:30 either, is it?

Did you notice that the minute hand is pointing to the 1?

Careful!

When the minute hand is pointing at a number remember that the numbers show 5 minutes for each one.

So when the minute hand points to the 1 the time is 5 minutes past the hour. The time shown on the clock above is 8:05 or 5 minutes past 8.

When the minute hand points to the 4, the time is 20 minutes past the hour or 4 times 5 minutes. When you were younger you learned to count by fives. You could find this time by counting by fives as you touched each number.

Let's Try It!

Set #3

What time is shown by each of these clocks?

Think About It!

How many minutes are left in an hour when the minute hand is pointing to the 11?

Answers are on page 171.

TIME AS A FRACTION

Did you know that fractions can be used to tell time? Pethagerus is looking at a clock that shows 4:15.

Now look at this clock again with the first 15 minutes shaded.

Do you see that one fourth of the clock is shaded? We can say that the time is quarter past the hour or quarter past 4.

If the time shown is 4:30, the clock is shaded to show 30 minutes have passed.

Do you see that one half of the clock is shaded? We can say that the time is half past the hour or half past 4.

As the hour goes by, the clock soon looks like this.

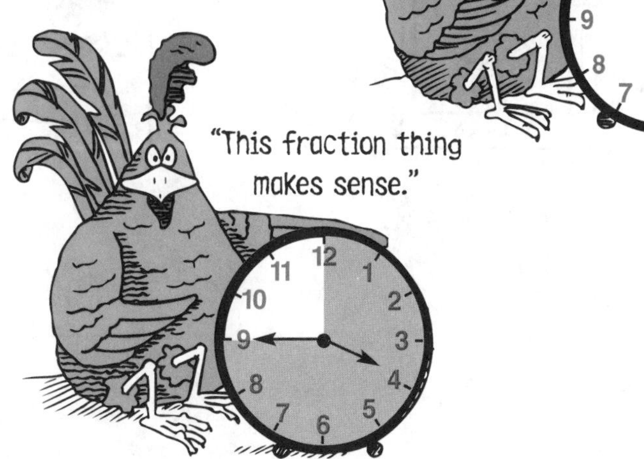

Careful!
We usually do not say that it is three quarters past the hour. Instead, we say the time is now a quarter to the next hour or a quarter to 5.

A.M. AND P.M.

Since our clocks only show 12 hours, and there are 24 hours in a day, we need to have a way to show which 12 hours we mean. We use A.M. and P.M. for that.

Watch Pethagerus as he goes through his day using A.M. and P.M.

A.M.

A.M.

P.M.

P.M.

Let's Try It! Set #4

Write A.M. or P.M. next to each time under the pictures.

Think About It!

At what time does A.M. change to P.M., and what is another name for that time? Draw the hands on the clock to show that time.

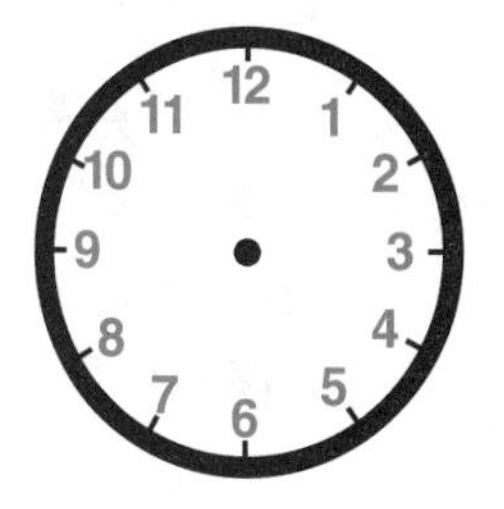

Answers are on pages 171–172.

WHAT TIME IS IT NOW?

Many of us have **digital clocks** in our homes. Did you know that each digital time has a matching time on a regular or **analog clock**?

Pethagerus has mixed up the clocks below. See if you can help him match the digital times with the analog clocks. Draw a line from each analog clock to its matching digital clock.

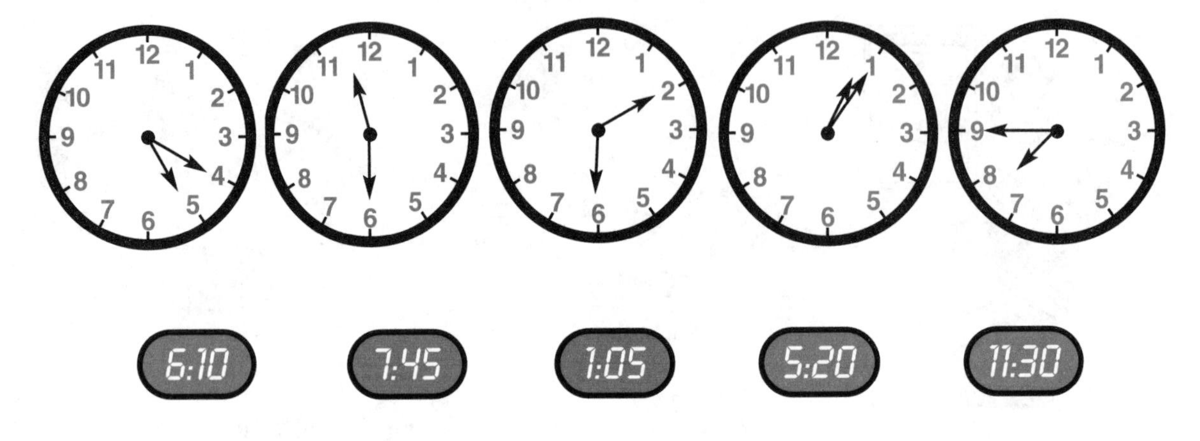

Can you tell if it is A.M. or P.M. by just looking at the clock?

On an analog clock, you can't tell if it is A.M. or P.M., but on digital clocks there is a little A or P to help you.

CALENDAR

Longer periods of time are measured by calendars. Calendars are used to keep track of days, weeks, months, and years. They are made in all sizes, and some have beautiful pictures to show the season for each month. Almost every home has a calendar handy to help family members remember important dates. Have you ever circled your birthday on a calendar?

Important!

1. There are 7 days in a week.
2. There are 12 months in a year.
3. There are about 52 weeks in a year.
4. There are 10 years in a decade.
5. There are 100 years in a century.

Below you see calendars for the months of July and August.

JULY

SUN	MON	TUES	WED	THUR	FRI	SAT
						1
2	3	4	5	6	7	8
9	10	11	12	13	14	15
16	17	18	19	20	21	22
23	24	25	26	27	28	29
30	31					

AUGUST

SUN	MON	TUES	WED	THUR	FRI	SAT
		1	2	3	4	5
6	7	8	9	10	11	12
13	14	15	16	17	18	19
20	21	22	23	24	25	26
27	28	29	30	31		

Matt and his family are going on a trip. They are leaving on July 17 and will return on August 5. How long are they going to be away?

First look at July 17 and shade it in. Now count down two Mondays until you come to July 31. That makes 2 weeks. Now add the first 5 days of August and you have 2 weeks and 5 days.

You can also use a calendar to plan your special days. Suppose your birthday is July 12. You want to invite your friends one week before that. What day of the week will that be? What date will that be?

Start at your birthday (July 12). Put your finger on it, and move it up one row. Your finger should be on Wednesday, July 5. That is when you should invite your friends. A calendar is set up so you don't have to count back 7 days to get a week. The days are lined up under the names of the days of the week.

Careful!

Not all calendars have the days of the week listed.

Some have just the first letter above the column for the days. In this case, usually Thursday is shown as "Th" so you don't confuse it with Tuesday. In some cases, the names of the days don't appear at all!

Let's Try It!

Set #5

Use this calendar of the month of May to answer your questions.

MAY

SUN	MON	TUES	WED	THUR	FRI	SAT
						1
2	3	4	5	6	7	8
9	10	11	12	13	14	15
16	17	18	19	20	21	22
23	24	25	26	27	28	29
30	31					

1. What day of the week is May 9th? ___________
2. How many Saturdays are in May? ___________
3. How many Tuesdays are in May? ___________
4. How many weeks and days are there from May 9th to May 25th? ___________
5. What day of the week is the first day of June in this year shown? ___________

Think About It!

Notice that there are five Sundays in the month of May. Can there ever be more than five of one day of the week in a month? Why? Or why not?

Answers are on page 172.

ELAPSED TIME

Elapsed time is the time that it takes to do something. For instance, if a birthday party begins at 1 P.M. and ends at 4 P.M. the elapsed time for the party is 3 hours.

If you live in an area that gets snow in the winter you might know a lot about elapsed time. The radio announcer says that your school will be opening two hours late today because of a snowstorm. Your school normally begins at 8:35 A.M. What time will it be opening today? Draw the hands on the clock on the right to show today's opening time.

Sometimes elapsed time does not work out on the hour. For instance, a movie begins at 11:15 A.M. and ends at 1:45 P.M. How long was the movie?

Count the hours first. From 11:15 A.M. to 12:15 P.M. is 1 hour. From 12:15 P.M. to 1:15 P.M. is another hour, and from 1:15 P.M. to 1:45 P.M. is 30 minutes. The total elapsed time is 2 hours and 30 minutes.

There are times when elapsed time is less than one hour. For instance, your math class might begin at 10:10 A.M. and end at 10:50 A.M. How long is your math class? The minute hand is on the 2 at 10:10 and is on the 10 at 10:50. Counting by fives from 2 to 10 is 8 fives or 40 minutes.

Let's Try It!

Set #6

Figure the elapsed time for each of these. Use the clock drawn here to help you.

1. Start time 8:40 and end time 9:20 ______________
2. Start time 4:10 and end time 7:35 ______________
3. Start time 11:30 A.M. and end time 2:15 P.M. __________
4. Start time 3:45 and end time 5:50 ___________
5. Start time 12:05 P.M. and end time 1:00 P.M. __________

Think About It!

Why is it important to use A.M. and P.M.?

Answers are on page 172.

Chapter 2

Does It Make Sense?

A Chapter About Place Value and Number Sense

Terms and Definitions

Digits: The symbols we use to make numbers: 0, 1, 2, 3, 4, 5, 6, 7, 8, 9.

Place value: The value of each digit in a number, based on its location in the number.

Period: Each group of three digits in a number.

Standard form: A number written in digits. Example: 398

Expanded form: A number written as the sum of each place value. Example: 300 + 90 + 8

Rounding: Replacing one number with another number that tells about how many. This is used most often to estimate.

IT'S THE WHOLE PICTURE—WRITING WHOLE NUMBERS TO 1,000

Did you ever hear the old saying "a place for everything and everything in its place?" Numbers work that way, too.

Before we start to talk about numbers, Pethagerus would like to introduce you to some helpful words.

Look at the number Pethagerus is reading. There are three symbols used to make it. They are called **digits**.

Each digit has a value called **place value**.

When three digits are placed together in a number they fall into a **period** on a place value chart.

Numbers that are written like the number Pethagerus is reading are written in **standard form**.

Numbers that are written as the sum of the values of the digits are written in **expanded form**.

Let's see if we can put some of these things to work. Look at Pethagerus' number again. To read this number aloud you say, "one hundred fifty-eight."

Careful!

Make sure that you never say the word *and* while reading whole numbers. We'll talk more about this later in Chapter 6.

Did you hear yourself say the value of the 1 when you read it? You said, "one hundred fifty-eight." The 1 is in the hundreds place and has a value of 100.

The value of the 5 is not so easy at first but you will get the hang of it. The 5 is in the tens place. Think of counting by tens. Five tens is 50, so the 5 in the tens place has a value of 50.

That leaves the ones place. No matter how big or small a number is there is always a ones place. In Pethagerus' number there is an 8 in the ones place. It has a value of 8.

Let's Try It!

Set #1

What is the value of each digit in the following numbers?

1. 392 ________________
2. 479 ________________
3. 581 ________________
4. 635 ________________
5. 247 ________________

Think About It!
What happens when there is a 0 in the tens or ones place?

Answers are on pages 172–173.

IS THIS THE RIGHT ORDER?

When you were younger you learned to count to ten and then to 100. You learned that it was important to get numbers in the proper order.

With three-digit numbers it is just as important to get them in the proper order. Pethagerus is here to help us get that order right.

Look at the hundreds place first. Do you see a different digit in each number? Good! Then this will be easy! The first number in order will be the one with the smallest hundreds place. The next one in order will be the medium or middle number in the hundreds place, and the last one will be the greatest number in the hundreds place. Let's see if Pethagerus has got them right.

What if the hundreds place has the same digit? Look at Pethagerus' cards now!

Now we must look at the tens place. Is there a different digit in each number in the tens place? Good! Since 3 is the lowest number, 638 would come first, then 5 would be next with 654. Finally, 672 would be last because 7 is the greatest number.

HOW DOES MY NUMBER COMPARE WITH YOURS?

Sometimes it is important to know how numbers compare in size. For instance, you and your friend might have collections of baseball cards. You can keep track of who has more cards by knowing how to compare numbers.

If you have 398 cards and your friend has 389 cards, who has more? The hundreds places are the same but the tens places are different. You have more cards because there is a 9 in the tens place of 398 and an 8 in the tens place of 389.

Careful!

Even though the same digits are in both numbers the value is different.

It is a good idea to use expanded form to help us with comparing numbers. To write 398 in expanded form think of the value of each place. The 3 is worth 300, the 9 has a value of 90, and the 8 is worth 8. Expanded form for 398 is 300 + 90 + 8. Expanded form for 389 is 300 + 80 + 9. Since 90 is greater than 80, 398 is greater than 389. We can write it like this:

398 > 389.

Careful!

The greater than > and less than < signs are easy to mix up. Remember that the less than sign looks like a sideways L.

Let's Try It!

Set #2

Compare these numbers. Write <, >, or = between them to make a correct math statement.

1. 495 ____ 594
2. 301 ____ 298
3. 732 ____ 742
4. 900 ____ 899
5. 678 ____ 678

Think About It!

Is the ones place ever important in ordering numbers?

Answers are on page 173.

HOW BIG IS ONE THOUSAND?

Now that we are comfortable with three-digit numbers, let's look at some bigger ones. Here is where we get to talk about periods of numbers. Look at the place value chart below. You can see that there are three groups of three numbers. Going from the right to the left, the first group is called the ones period. The middle one is the thousands period and the last one on the left is the millions period.

Millions Period			Thousands Period			Ones Period		
hundreds	tens	ones	hundreds	tens	ones	hundreds	tens	ones

Pethagerus has a helpful hint for you.

The name for each period is the same as the name of the ones place in it.

For example: The ones place in the thousands period is called the thousands place. The ones place in the millions period is called the millions place.

Have you ever thought about how big one thousand is? Think of marbles in a container. Each container has 10 marbles in it.

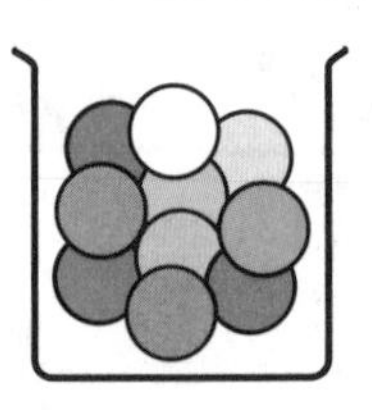

If you put 10 of those containers together in a box you have 100 marbles. Now take 10 of those boxes of 100 marbles and pile them up. You have just put together 1,000 marbles!

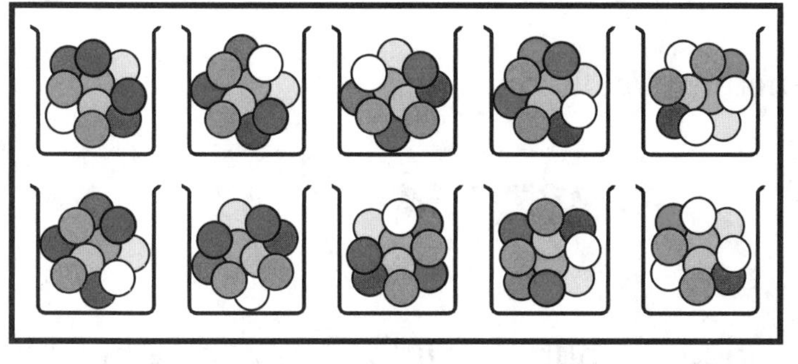

Pethagerus seems to think that 1,000 marbles are a lot! Can you think of anything that you have seen like that? Do you think a tree might have 1,000 leaves on it? Is it possible to count them all? Do you think a taller tree might have more leaves on it than a shorter one? Have you ever seen 1,000 people anywhere? Sometimes, at the beginning of a baseball or football game, you will hear the announcer say how many thousand people are attending the game.

Let's think about numbers in the thousands period. Here is a number to look at.

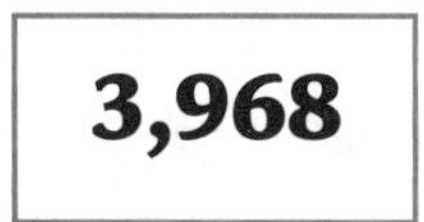

Read that number aloud to yourself. Did you say, "three thousand nine hundred sixty-eight"? You were right!

Try this one.

43,521

Did you say "forty-three thousand five hundred twenty-one"? Good!

Now let's go for something really big.

987,654

Try to read this one aloud to yourself. Did you say "nine hundred eighty-seven thousand, six hundred fifty-four"? Now you are ready to try some more!

Let's Try It!

Set #3

Read these numbers aloud to yourself.

1. 5,952
2. 67,432
3. 479,025
4. 9,999
5. 824,563

Think About It!

What really big number comes right after 999,999?

Answers are on page 173.

HOW BIG IS ONE MILLION?

If you knew the answer to the last Thinker's Question, then you will like what comes next. Have you ever thought about how big a million is? Let's go back to the marbles. Remember the 10 boxes of 100 marbles we piled up to get 1,000 marbles? Imagine 1,000 of those piles of 1,000 marbles! That is what it would take to get one million!

Can you think of a million of anything? Have you looked at the sky at night? There are at least a million stars for us to see. Can you count them? What do you think it would be like to count to 1,000,000? How long would it take? How long would it take to write one million letters in your notebook? Would it fill the notebook or more?

GETTING 'ROUND TO IT—ROUNDING NUMBERS

Sometimes it is not necessary to know the exact value of a number. When that happens we round a number.

No, we are not drawing circles around the number! We are really changing the number slightly in order to use it for some problems. Actually, to round a number means that we will find the nearest value of a number based on a given place value.

Let's look at rounding to the nearest 10 first. Study the number line below.

Write 33 on the number line in its proper place. Is 33 closer to 30 or to 40? Clearly, it is closer to 30, so 33 rounded to the nearest ten is 30. Now write 38 on this number line. Is 38 closer to 30 or to 40? You were right if you said 40, so 38 rounded to the nearest ten is 40. Now write 35 on this number line in its proper place. What is special about 35? It is halfway between 30 and 40, so how do we round 35 to the nearest ten? This was a problem for mathematicians until it was decided that all numbers in the middle would round up. That means 35 rounded to the nearest ten is 40.

Larger numbers can also be rounded to the nearest ten. Most of the time, the digits in the higher places just come along for the ride. Look at the number line below.

Find the place for 864, and write that number on the number line. Is 864 nearer to 860 or 870? It is nearer to 860, so 864 rounded to the nearest ten is 860. Now find 867, and write it on the number line. Is 867 nearer to 860 or 870? You were right if you said 870, so 867 rounded to the nearest ten is 870. How are you doing so far?

Now let's look at what happens sometimes to rounding larger numbers.

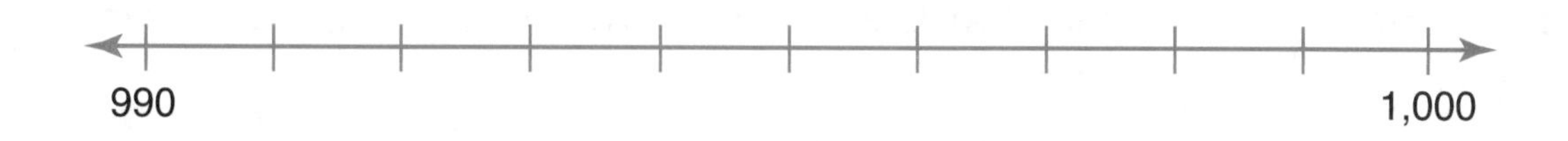

Write 992 on this number line in its proper place. Is 992 nearer to 990 or 1,000? It is nearer to 990, so 992 rounded to the nearest ten is 990. Now write 998 on the number line in its proper place. Is 998 nearer to 990 or 1,000. You were right if you said 1,000, so 998 rounded to the nearest ten is 1,000! In this case, all the digits changed to write the nearest ten.

Let's Try It!

Set #4

Round the following numbers to the nearest ten.

1. 56 ________
2. 25 ________
3. 378 ________
4. 1,541 ________
5. 89,659 ________

Think About It!

Is 1,000 really a ten? Why or why not?

Answers are on pages 174–175.

WET NOODLE NUMBER LINE

Have you ever had a wet noodle left on your plate at the end of a meal? Have you noticed that they are not straight but wavy like this drawing?

Imagine that this wet noodle is a number line like the one below.

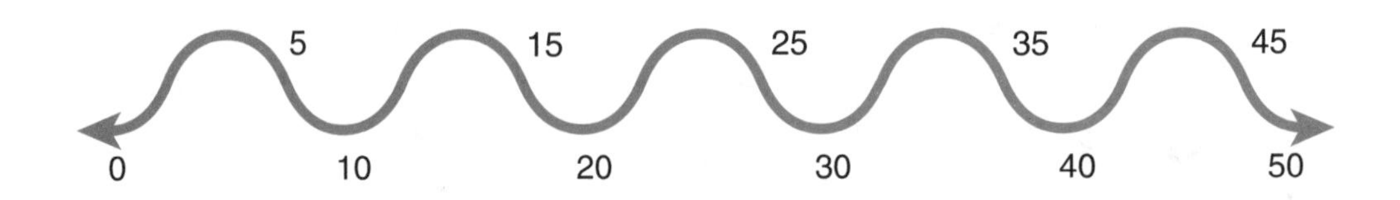

Do you see the numbers 0, 10, 20, 30, 40, and 50 at the bottoms of the valleys formed by the noodle? Now look at the tops of the peaks. Do you see that the numbers with 5 in the ones place are leaning to the right? That will help you to use this wet noodle number line to round numbers to the nearest ten. Find the number 23 that is also placed on this special number line. What would happen if this number rolled down the hill? It would land in the valley of 20, so 23 rounded to the nearest ten is 20. Let's try another. Write the number 8 on this number line. If 8 rolled down the hill where would it be? It would be in the valley of 10, so 8 rounded to the nearest ten is 10. Look at the number 15. What would happen to 15 if it rolled down the hill? It would land in the valley of 20, so 15 rounded to the nearest ten is 20. The numbers 5, 15, 25, 35, and 45 all lean to the right to help you remember that those numbers round up to the next ten.

Look at the new wet noodle number line drawn for you below.

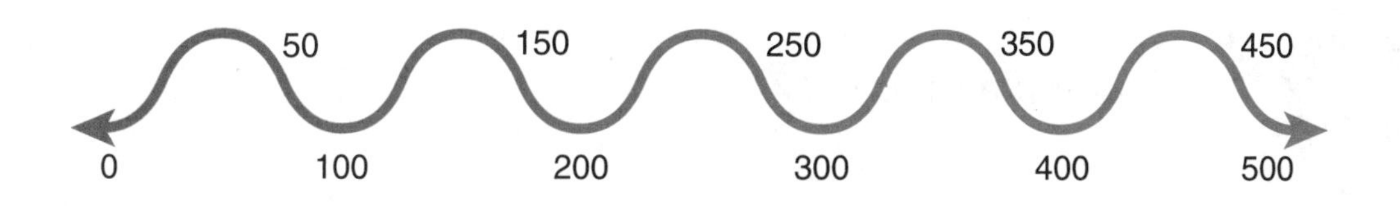

This number line will help you to round numbers to the nearest hundred. Do you see the number 328 already placed on the number line? What would happen to 328 if it rolled down the hill? It would land in the valley of 300, so 328 rounded to the nearest hundred is 300. Let's try this one. Write the number 163 on the number line. What valley does 163 roll down to? You were right if you said 200. We see that 163 rounded to the nearest hundred is 200. What about the number 350? That is already on the number line and leaning to the right. What valley will it land in? It will land in the valley of 400 because 350 rounded to the nearest hundred is 400.

The same rules apply when rounding larger numbers to the nearest hundred. For instance, if you want to round 1,298 to the nearest hundred, write 298 on the number line. 298 rounded to the nearest hundred is 300, so 1,298 rounded to the nearest hundred is 1,300.

You can draw your own wet noodle number lines when you need to round numbers.

Let's Try It! Set #5

Round the following numbers to the nearest hundred.

1. 129 ________
2. 451 ________
3. 764 ____________
4. 3,897 ______________
5. 15,678 ________________

Think About It!

What is 49 rounded to the nearest hundred?

Answers are on page 174.

When do you need to round numbers? Look for the magic word *about*. Here is an example. Mrs. Kushner's class put on a play that was attended by 158 people. About how many people came to the play? The answer could be about 160 people if you rounded to the nearest ten and about 200 people if you rounded to the nearest hundred.

You will most often see rounding take place when someone is talking about the attendance at a popular sporting event like baseball, football, or the Olympics. You will see a newspaper headline something like this:

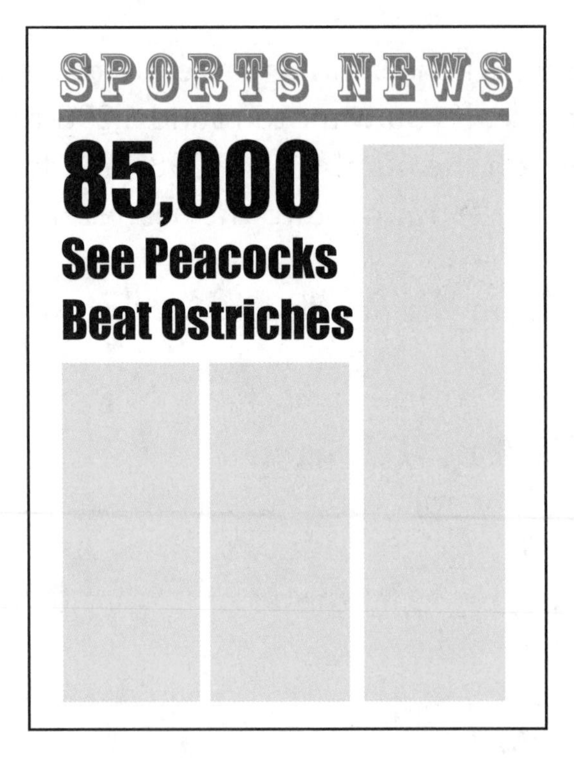
SPORTS NEWS

85,000
See Peacocks
Beat Ostriches

Chapter 3
Does It Add Up?
A Chapter About Addition
10
+3

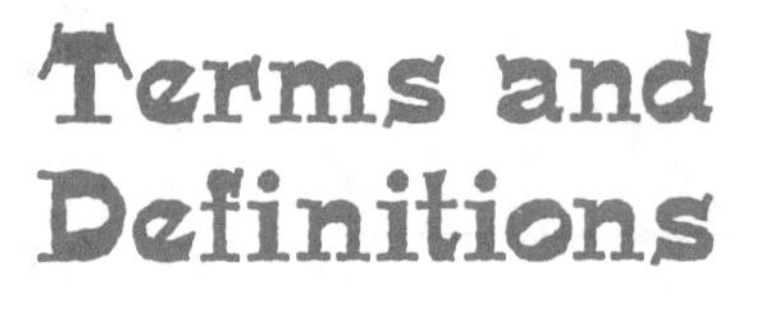

Terms and Definitions

Addends: The numbers added together in an addition problem.

Sum: The answer to an addition problem.

Commutative property: This property states that the order of the addends does not change the sum.

Associative property: Sometimes called the grouping property, this property states that the grouping of the addends does not change the sum.

Identity property: This property states that when zero is added to a number the sum is that number.

WHAT HAPPENS TO NUMBERS WHEN YOU ADD THEM?

Do you remember the first operation you learned when you were younger? The chances are that you learned addition first. Did you ever think about what happens to numbers when you add them? Pethagerus likes cookies so let's try it.

If Pethagerus puts all the cookies on one plate he is adding them together. The math sentence or equation looks like this:

$3 + 4 = 7$

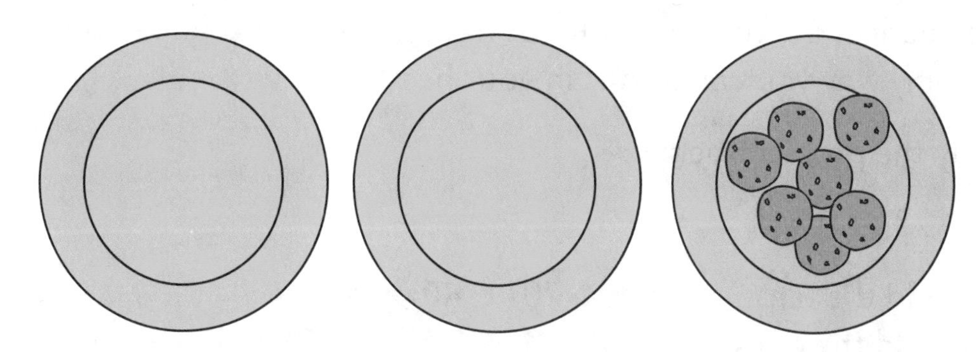

Now Pethagerus has seven cookies on the plate. Seven is a larger number than three or four, so adding numbers results in a larger number.

Do you think this always happens? Study the equations below. Remember: The numbers that you add together are called **addends** and the answer is called the **sum**.

$$6 + 2 = 8$$
$$5 + 4 = 9$$
$$3 + 8 = 11$$
$$12 + 7 = 19$$

Did you check the sums? Are they all larger than each of the addends? You were right if you said yes.

What about some larger numbers? Here they are!

$$23 + 64 = 87$$
$$51 + 75 = 126$$
$$95 + 64 = 159$$
$$99 + 87 = 186$$

Did you notice that each time the sum is a larger number than either of the addends?

MENTAL ADDITION

As you get older you will do more and more mental math. This often happens when you are not in school and need to add numbers quickly. The chances are that you will not have a pencil and paper with you. The only way you can add is in your head.

Study the problem below:

$$73 + 42$$

First add the tens.	$70 + 40 = 110$
Then add the ones.	$3 + 2 = 5$
Add those two sums.	$110 + 5 = 115$

Let's try adding numbers mentally in the hundreds.

Here's the problem:

$264 + 415$

First add the hundreds.	$200 + 400 = 600$
Then add the tens.	$60 + 10 = 70$
Now add the ones.	$4 + 5 = 9$
Add to find the final sum.	$600 + 70 + 9 = 679$

Did you notice that the final sum looks a lot like the expanded form that we explored in Chapter 2?

Let's Try It! Set #1

Add the following numbers in your head.

1. $52 + 63$
2. $38 + 51$
3. $55 + 73$
4. $281 + 517$
5. $327 + 662$

Think About It!

How do you add a larger number in your head with a smaller number?

Example: $326 + 52$

Answers are on page 174.

PROPERTIES OF ADDITION

Have you heard about the properties of addition? Perhaps your teacher has mentioned them to you in school. The properties are rules that help you. Let's look at them to see how they can help!

The first property of addition we will explore is the **commutative property**. This property states that the order of the addends does not change the sum. Here's how it works:

$$3 + 4 = 7$$
$$4 + 3 = 7$$
So $3 + 4 = 4 + 3$
And $67 + 89 = 89 + 67$

The next property is called the **grouping property**. It is also called the **associative property**. This property states that the grouping of the addends does not change the sum. Here's how this one works:

$$2 + (4 + 6) = (2 + 4) + 6$$
$$2 + 10 = 6 + 6$$
$$12 = 12$$

Careful!
Add the numbers in ( ) first!

The last property we will look into is called the **identity property**. This property states that when zero is added to a number, the sum is that number. For example:

$$8 + 0 = 8$$
$$0 + 23 = 23$$

It might help you to remember this property's name by thinking of the word *identical*. A number is identical to the sum of itself and zero.

Let's Try It!

Set #2

Solve these problems using the properties we just talked about. After you have solved the problem, place a C for commutative, I for identity, or A for associative or grouping property on the other line provided.

1. 59 + 0 = _____ _____
2. 34 + 62 = _____ + 34 _____
3. 0 + 398 = _____ _____
4. (3 + 5) + 2 = 3 + (_____ + 2) _____
5. 857 + _____ = 245 + 857 _____

Think About It!

Is it possible to use more than one property in a single equation (number sentence)?

Answers are on page 175.

MISSING ADDENDS

Do you remember when you were younger and your teacher talked about missing addends? Did you wonder if she ever found them?

Actually, figuring missing addends is a good exercise to use to begin to learn to subtract. Here's how it works.

Find the missing addend in the following equation:

$$7 + \_\_\_\_ = 15$$

You can solve this problem by trying different numbers until one works. This method is called guess and check because you try a number and check to see if it works. If it doesn't work, then you try another number using what you have learned from the previous guess. Look up this method in Chapter 12.

You can also use subtraction to help you here. Since 15 is the larger number, you can subtract 7 from it so it looks like this:

$$15 - 7 = 8$$

Now let's go back to the first problem and try 8 as the missing addend.

$$7 + 8 = 15$$

It works because addition and subtraction are friendly to each other in a special way. Check this out in Chapter 4.

WAYS TO ADD

There are several ways to add numbers together. As you get older you will find that the way you most often use is by mental math. We talked about that earlier in this chapter.

You will also use a calculator as you get older. This may sound like cheating to some of your parents, but knowing good calculator skills will become important to you in your later math studies and also later in your life!

The way you add most often at the present time is the paper-and-pencil method. Let's see how that works!

Step One:

```
  368
  952        Add the numbers in the ones column.
 +401
 ----
    1              11 = 1 ten  1 one
```

Step Two:

```
   1
  368
  952        Add all the numbers in the tens column.
 +401
 ----
   2         120 = 1 hundred 2 tens
```

Step Three:

```
  1
  368
  952        Add all the numbers in the hundreds column.
 +401
 ----
 1721
```

If you are like most students, you like to add.

Let's Try It!

Set #3

1. $$\begin{array}{r} 5{,}793 \\ +7{,}351 \\ \hline \end{array}$$

2. $$\begin{array}{r} 459 \\ 731 \\ +865 \\ \hline \end{array}$$

3. $$\begin{array}{r} 9{,}834 \\ +8{,}609 \\ \hline \end{array}$$

4. $$\begin{array}{r} 693 \\ 452 \\ +975 \\ \hline \end{array}$$

5. $$\begin{array}{r} 6{,}832 \\ 7{,}895 \\ +5{,}649 \\ \hline \end{array}$$

Think About It!

What happens to the value of numbers when you add them?

Answers are on page 175.

Chapter 4

Do You Say, “Take Away”?

A Chapter About Subtraction

Terms and Definitions

Difference: The answer to a subtraction problem.

Fact Family: A set of related addition/subtraction number sentences.

WHEN AM I GOING TO SUBTRACT?

Do you remember the first time you heard someone say "take away" to you? The chances are that you were in a classroom. Many teachers use those words to remind their students to subtract. However, subtraction is not only about taking away. If you have any brothers or sisters you have most likely heard someone ask you, "What is the difference in your age and your brother's or sister's?" The magic word in that sentence is *difference*. The answer to a subtraction problem is called the **difference**!

What does that mean about subtraction? Actually, subtraction is both taking away and comparing two numbers. Let's see how that works.

Here is a subtraction problem to work with:

$$9 - 3 = 6$$

What does the problem mean? In the case of you and your siblings it could mean this: You are 9 years old and your brother is 3 years old. What is the difference in your ages? The difference is 6 years. If this were strictly a "take away" problem we might be looking at something like this: Drew had nine cookies and ate three of them. How many cookies does he have left? He has six cookies.

Mental Subtraction

The use of mental math is not as friendly to subtraction as it is to addition and multiplication. It is really best to tackle this when the numbers work well. For example, you may need to use mental math with money but stay with numbers that don't need regrouping. Here's how it works:

$8.50 - $5.30

First, subtract the cents. 50¢ - 30¢ = 20¢
Then subtract the dollars. $8 - $5 = $3
Put that together and you have $3.20.

Smaller Digits on Top

Now that we have warmed up a bit, let's tackle some of the really good stuff about subtraction.

Let's try some subtraction with smaller digits on the top. Here's how it works:

$$\begin{array}{r} 932 \\ -585 \\ \hline \end{array}$$

Begin with the ones place.
2 < 5 so we will regroup.

$$\begin{array}{rrr} & 2 & 12 \\ 9 & \not{3} & \not{2} \\ -5 & 8 & 5 \\ \hline & & 7 \end{array}$$

Now you can subtract
12 minus 5 to get 7.

$$\begin{array}{rrr} 8 & 12 & 12 \\ \cancel{9} & \cancel{3} & \cancel{2} \\ -5 & 8 & 5 \\ \hline & 4 & 7 \end{array}$$

In the tens place,
2 < 8 so we will regroup.
12 minus 8 is 4.

$$\begin{array}{rrr} 8 & 12 & 12 \\ \cancel{9} & \cancel{3} & \cancel{2} \\ -5 & 8 & 5 \\ \hline 3 & 4 & 7 \end{array}$$

In the hundreds place, we subtract
8 minus 5 to get 3.

Careful!

Make sure that you cross out the number you are regrouping. When there is a lot of regrouping in a problem, check your answer to make sure it makes sense. You can also check your answer by adding the two smallest numbers in the problem. If you are right, their sum should be the largest number.

Let's Try It!

Set #1

Find the difference.

1. $\begin{array}{r} 831 \\ -576 \\ \hline \end{array}$

2. $\begin{array}{r} 956 \\ -395 \\ \hline \end{array}$

3. $\begin{array}{r} 674 \\ -485 \\ \hline \end{array}$

4. $\begin{array}{r} 743 \\ -266 \\ \hline \end{array}$

5. $\begin{array}{r} 572 \\ -181 \\ \hline \end{array}$

Think About It!

What happens to numbers that are subtracted from each other?

Answers are on page 176.

LET’S THROW A ZERO IN THE TOP NUMBER!

How are you doing so far?

Now let’s try something a little different. What do you think happens when you have a zero in the top number?

Here’s how it works:

$$\begin{array}{rrr} & 7 & 10 \\ 5 & \cancel{8} & \cancel{0} \\ -3 & 7 & 2 \\ \hline \end{array}$$

$0 < 2$ so regroup to make $10 - 2$.
Now subtract as usual.

$$\begin{array}{rrr} & 7 & 10 \\ 5 & \cancel{8} & \cancel{0} \\ -3 & 7 & 2 \\ \hline 2 & 0 & 8 \end{array}$$

What if the zero is in the tens place? Let's see how that works.

$$\begin{array}{rrr} & 9 & \\ 6 & \cancel{10} & 14 \\ \cancel{7} & \cancel{0} & \cancel{4} \\ -3 & 8 & 6 \\ \hline \end{array}$$

Let's see. I know that 4 < 6, but when I go to regroup in the tens there is nobody home. So I regroup in the hundreds, regroup the tens again, and I'm ready to subtract.

$$\begin{array}{rrr} & 9 & \\ 6 & \cancel{10} & 14 \\ \cancel{7} & \cancel{0} & \cancel{4} \\ -3 & 8 & 6 \\ \hline 3 & 1 & 8 \end{array}$$

Let's Try It!

Set #2

Find the difference.

1. $\begin{array}{r} 906 \\ -598 \\ \hline \end{array}$

2. $\begin{array}{r} 801 \\ -273 \\ \hline \end{array}$

3. $\begin{array}{r} 403 \\ -284 \\ \hline \end{array}$

4. $\begin{array}{r} 604 \\ -385 \\ \hline \end{array}$

5. $\begin{array}{r} 505 \\ -467 \\ \hline \end{array}$

Think About It!

What happens if both the ones and tens places have zeroes in the top number?

Answers are on pages 176–177.

FAMILY OF FACTS—ADDITION AND SUBTRACTION

In Chapter 3 during the talk about missing addends, there were hints about how friendly addition and subtraction are. Did you think about that? If you are like most students, you would rather add than subtract, but you can put them both together to make a family of facts.

Watch this!

If you know these facts:

$$5 + 3 = 8$$

and

$$3 + 5 = 8$$

Then you also know:

$$8 - 5 = 3$$

and

$$8 - 3 = 5$$

Altogether, this is a **family of facts** or **fact family**.
Here's another one.

Use 2, 4, and 6.

$$2 + 4 = 6$$
$$4 + 2 = 6$$
$$6 - 4 = 2$$
$$6 - 2 = 4$$

Let's Try It!

Set #3

Use these sets of numbers to make fact families.

1. 7, 8, 15
2. 9, 4, 13
3. 3, 7, 10
4. 5, 4, 9
5. 11, 9, 20

Think About It!

What happens to addition/subtraction fact families that contain doubles? For example; 4, 4, 8.

Answers are on page 177.

Chapter 5
HOW MANY GROUPS?
ARE THERE ANY
LEFTOVERS?
A Chapter About
Multiplication and Division
Le Coop
3+5=8
8
-5
3
8
-3
5
5+3=8

Terms and Definitions

Array: An arrangement of objects or pictures in rows and columns.

Factors: The numbers being multiplied.

Product: The answer to a multiplication problem.

Commutative property: Sometimes called the order property, this property states that the order of the factors does not change the product.

Associative property: Sometimes called the grouping property, this property states that the grouping of the factors does not change the product.

Identity property: This property states that any number multiplied by one stays the same.

Zero property: Any time you multiply by zero the product will be zero.

Dividend: The number that is to be divided.

Divisor: The number that divides the dividend.

Quotient: The answer to a division problem.

Prime number: A number with exactly two factors, itself and one.

Composite number: A number with more than two factors.

Prime factorization: Method where you determine all factors of a number that are prime factors. Example: The prime factors of 24 are $2^3 \times 3$.

Multiples: These are products of the same number. For example: Multiples of 3 are 3, 6, 9, 12, 15, and so on. Any product of 3 is a multiple of 3.

WHAT HAPPENS TO NUMBERS WHEN YOU MULTIPLY?

The moment has arrived. Your teacher is about to hand out the dreaded test and everyone in your classroom is nervous. She takes out a box of pencils and says that each of you will get two, but she only has 40 pencils. There are 22 students in your class. Are there enough for everyone to get two pencils? How do you solve this problem? One way would be to use **multiplication**. This operation is so much fun because it is a fast way to add. In this problem, for example, you could add two 22 times to see how many pencils your class needs, or you could simply multiply 22 times 2 to get 44. It looks as if not everyone in your class will be able to get two pencils. There are not enough!

So just what happens to numbers when you multiply them? Let's see:

$3 \times 4 = 12$
$5 \times 2 = 10$
$6 \times 3 = 18$

Every time we multiplied, we ended up with bigger numbers! Do you think that always happens? Let's review this idea at the end of the chapter. See you there!

OTHER WAYS TO MULTIPLY

Did you know that there are two ways to say a multiplication equation? Let's look at this one:

$4 \times 5 = 20$

The first way to read this equation is to say, "Four times five equals twenty."

You can also say, "Four groups of five equal twenty." You can draw a picture of this one.

There is another kind of picture called an **array** that can be drawn to show a multiplication problem. An array shows pictures in rows and columns. Look at this one:

This array shows three rows of two apples in each row and can be written as $3 \times 2 = 6$.

It also shows two columns of three apples in each column which can be written as $2 \times 3 = 6$.

As a matter of fact, drawing a picture is just one of the ways to help you multiply numbers. You can also skip count on a number line to find the **product** or answer to a multiplication problem. Look at the number line below to see how that works.

0 1 2 3 4 5 6 7 8 9 10 11 12 13 14 15 16 17 18 19 20

These arrows are skip counting by threes. They are also showing multiples of 3 every time they land on the number line. The number of arrows is one of the **factors**, the amount of numbers being skipped is the other factor, and the number it lands on is the **product**.

For example, the second arrow, or arrow number 2, lands on 6. That means $2 \times 3 = 6$.

As you practice more with your multiplication facts, you will realize that you are learning them by heart. It is a good idea to memorize them as soon as possible because your teachers will ask you to anyway!

Important!
Know the facts!

PROPERTIES OF MULTIPLICATION

The properties of multiplication will help you learn these facts. The first property is called the **commutative property**. Sometimes you might hear it called the **order property**. The commutative property states that the order of the factors does not change the product. Here is an example:

$3 \times 5 = 15$

$5 \times 3 = 15$
So $3 \times 5 = 5 \times 3$

In other words, when you learn all the 2 times table to 12 you have actually learned the first fact in the 12 times table!

$12 \times 2 = 2 \times 12$

The next property is called the **associative** or **grouping property**. This property states that the grouping of the factors does not change the product. Here is an example:

$(2 \times 3) \times 4 = 2 \times (3 \times 4)$

Let's check this to make sure.

$$
\begin{aligned}
(2 \times 3) \times 4 &= 2 \times (3 \times 4) \\
6 \times 4 &= 2 \times 12 \\
24 &= 24
\end{aligned}
$$

Now let's take a peek at the **identity property**. It states that any number multiplied by one remains the same. Here are some examples:

$7 \times 1 = 7$ and $43 \times 1 = 43$ and $1 \times 28 = 28$

The final property is called the **zero property**. It states that any time you multiply by 0 you get a product of 0. Here are some examples:

$0 \times 9 = 0$ and $15 \times 0 = 0$ and $392 \times 0 = 0$

Let's Try It!

Set #1

Put these properties to work by doing these problems. Fill in the missing number in the problem and then write the letter A for associative property, C for commutative property, I for identity property, or Z for zero property on the line after the problem.

1. $5 \times 3 = \_\_\_\_\_\_ \times 5$ ________
2. $2 \times (6 \times 1) = (\_\_\_\_\_\_ \times 6) \times 1$ ________
3. $98 \times 0 = \_\_\_\_\_\_$ ________
4. $23 \times 1 = \_\_\_\_\_\_$ ________
5. $\_\_\_\_\_\_ \times 7 = 7 \times 3$ ________

Think About It!

Is it possible to have more than one property in a single problem? If so, give an example.

Answers are on pages 177–178.

WHAT HAPPENS WHEN YOU DIVIDE?

If you have any brothers or sisters, you know a lot about this next topic. Have you ever been asked to share something with your family members? Chances are that you were **dividing**. Here's how it can happen. You have a nice collection of marbles and you want to play with them. Your two brothers also want to play marbles, but they don't have any. You have to divide the marbles in order to play with your brothers.

There are 24 marbles in your collection. Each of you will get the same number of marbles to play with. This is how it looks.

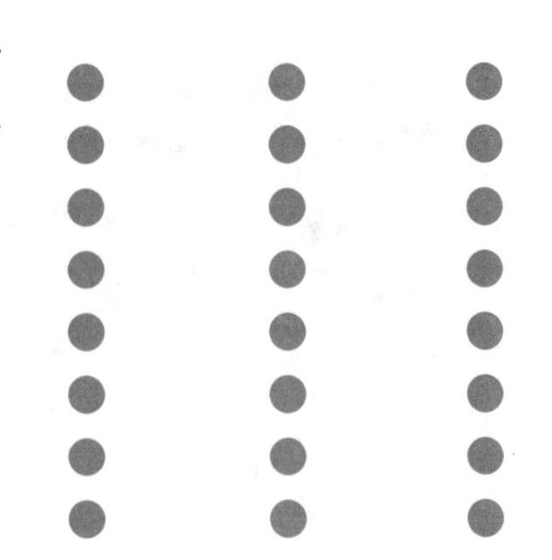

The division sentence will look like this:

$$24 \div 3 = 8$$

or like this:

$$3\overline{)24}^{\,8}$$

Did you know that each of these numbers has a name in a division problem? In this problem 24 is the **dividend**. Do you see the word *divide* in the word *dividend*? The dividend is almost always the largest number. The **divisor** is the number that does the work, the actual dividing. That is the number 3 in this problem. The **quotient** is the answer, in this case 8, to the problem.

What actually happens to numbers when you divide them? Let's look back for a moment. We started with 24 marbles and divided them into 3 groups with 8 marbles in each group. In other words, we started with 24 and ended with 8. Hmmmm . . . The quotient is smaller than the dividend. Let's try some more problems.

$$8 \div 2 = 4$$
$$10 \div 5 = 2$$
$$18 \div 6 = 3$$

OTHER WAYS TO DIVIDE

Do you remember what happened to numbers in Chapter 4 when we subtracted them? The difference was always a smaller number! Let's see if we can use some subtraction to help us with division!

We'll start with a simple problem:

$$6 \div 2 = 3$$

To use subtraction to do this problem, start with $6 - 2 = 4$. Then subtract $4 - 2 = 2$. And finally subtract $2 - 2 = 0$. Do you see that we had to subtract 2 three times to get back to 0? That is the same thing as dividing 6 by 2. We get a quotient of 3 with nothing left over.

Try this one:

$$20 \div 4 = 5$$

$$\begin{array}{r} 20 \\ -4 \\ \hline 16 \end{array} \qquad \begin{array}{r} 16 \\ -4 \\ \hline 12 \end{array} \qquad \begin{array}{r} 12 \\ -4 \\ \hline 8 \end{array} \qquad \begin{array}{r} 8 \\ -4 \\ \hline 4 \end{array} \qquad \begin{array}{r} 4 \\ -4 \\ \hline 0 \end{array}$$

It took five subtraction problems using four to get back to zero, so $20 \div 4 = 5$.

FAMILY OF FACTS— MULTIPLICATION AND DIVISION

In Chapter 3 with addition, the sum was always bigger. Now we are seeing that the product is always bigger when we multiply. We are also seeing that when we divide, the quotient is always smaller just as the difference is always smaller when we subtract. Do you remember the addition/subtraction families of facts we worked with in Chapter 4? There are families of facts with multiplication and division, too. Here's how it works:

$$3 \times 4 = 12$$
$$4 \times 3 = 12$$
$$12 \div 4 = 3$$
$$12 \div 3 = 4$$

Let's Try It!

Set #2

Make multiplication/division fact families using the three numbers given in each problem.

1. 3, 5, 15
2. 8, 2, 16
3. 6, 5, 30
4. 3, 7, 21
5. 4, 6, 24

Think About It!

See if you can make some multiplication/division fact families of your own. Check your work.

Answers are on page 178.

FACTORS AND MULTIPLES

Careful!

Now let's talk about factors and multiples. This is an area that sometimes causes confusion! Here's why: At the beginning of this chapter, we learned that factors are the numbers we multiply to get a product. Multiples are all the numbers that result when you count by a certain number.

Take a look:

In this problem

$$5 \times 4 = 20$$

the 5 and the 4 are factors.

Here are some multiples of 5:

5, 10, 15, 20, 25, 30, 35, . . .

Actually, these multiples continue on forever but factors are just the few numbers that appear in a multiplication equation.

PRIME AND COMPOSITE NUMBERS

Have you ever wondered what it must have been like to study math long, long ago? There were no calculators or computers, so there was a lot more work involved. Some of the best early mathematicians were Greek. One of them, named Eratosthenes, thought a lot about numbers and their factors. He discovered that there was a small group of numbers that had exactly two factors, the number itself and one. He called these numbers **prime numbers**. All the other numbers that have more than two factors he called **composite numbers**. By the way, one and zero are neither prime nor composite.

PRIME FACTORIZATION—WHAT'S UNDER YOUR PRIME FACTOR TREE?

Here is a list of all the prime numbers that are less than 100:

2, 3, 5, 7, 11, 13, 17, 19, 23, 29, 31, 37, 41,
43, 47, 53, 59, 61, 67, 71, 73, 79, 83, 89, 97

Now that you have this information, you can do something that is a lot of fun!

It's called **prime factorization**. Here's how is works:

Take a number like 24. Think of two factors for a product of 24. We will use 6 and 4 for this example. Now think of factors for 6 and 4. We will use 2 and 3 for 6 and 2 and 2 for 4. We now have the factors 2, 2, 2, and 3. Those are all prime numbers. Put them in a multiplication problem and they look like this:

$$2 \times 2 \times 2 \times 3 = 24$$

This process is most often shown in a prime factorization tree that looks like this:

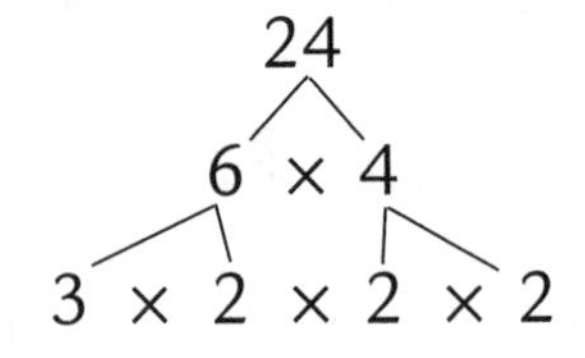

You can also write these prime factors using an exponent. It looks like this:

$$2^3 \times 3$$

Are there other factors that we could have begun our tree with for 24? How about 8 and 3? Let's see how that would work:

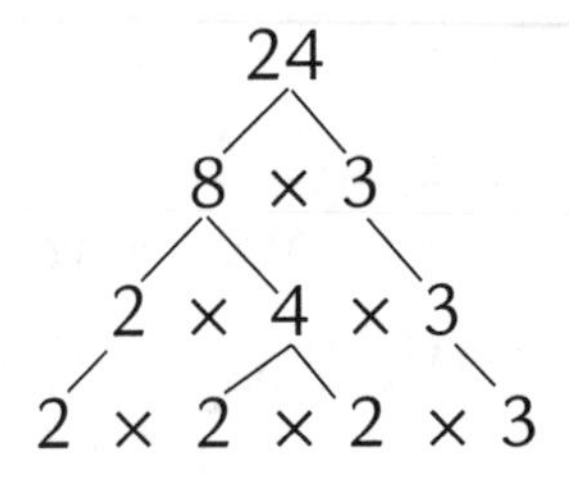

Let's Try It!

Set #3

Draw some prime factor trees for each of the products below.

1. 12
2. 25
3. 18
4. 30
5. 28

Think About It!

Does the prime factor tree get larger as the product gets larger?

Answers are on pages 178–179.

COMMON FACTORS

Have you ever been asked what you and your friend "have in common"? Did you ever wonder what that meant? In the study of math, when we talk about things in common, we mean things that are the same.

When we talk about common factors we are talking about the factors of two or more different numbers that are the same. Here's how it works:

The factors for 8 are 1, 2, 4, and 8.

The factors for 12 are 1, 2, 3, 4, 6, and 12.

The common factors for 8 and 12 are 2 and 4. We usually do not mention 1 because it is found in every set of factors.

Let's find the common factors for 12 and 18.

The factors for 12 are 1, 2, 3, 4, 6, and 12.
The factors for 18 are 1, 2, 3, 6, 9, and 18.
The common factors for 12 and 18 are 2, 3, and 6.

We will talk more about this when we think about fractions in Chapter 10.

COMMON MULTIPLES

Another time in math that we use the word *common* is when we think about common multiples. These are multiples that are found on the list of multiples for more than one number. For instance, the first ten multiples of 3 are 3, 6, 9, 12, 15, 18, 21, 24, 27, and 30. The first ten multiples of 4 are 4, 8, 12, 16, 20, 24, 28, 32, 36, and 40. The common multiple of 3 and 4 in this group are 12 and 24. You will think more about this when we work on fractions in Chapter 10.

Chapter 6
HOW MUCH DID YOU SPEND?
A Chapter About Money

Terms and Definitions

Equivalent sets: Sets of money with the same value but different coins.

Counting on: A method of making change by counting from the amount spent to the amount given to pay.

WHAT ARE THESE COINS WORTH?

Did you ever wonder where money comes from? Did you know that money is probably being made as you read this?

Pethagerus is thinking about money, too. Let's see what you remember. Look at the coins below and see if you can name them and identify their value. Write your answers on the lines below each coin.

Careful!
The dime is smaller in size than the nickel and the penny, but it has more value.

How did you do? Here are the answers to check your work.

Half-dollar	Quarter	Dime	Nickel	Penny
50 cents	25 cents	10 cents	5 cents	1 cent

CAN YOU MAKE EQUIVALENT SETS?

When you have more than one coin in your pocket, you can add their value to find out how much you have. Sometimes there is more than one combination of coins for the same value. These are called **equivalent sets**.

How many different ways are there to make 10¢?

There are four equivalent sets of coins to show 10¢.

Suppose you had 15¢ in your pocket. How many different sets of coins could you have? It's a good idea to start with the largest-value coins and work your way down to the pennies. That way you will be sure to get all the sets.

The largest-value coins you can use to make 15¢ are one dime and one nickel.

Now exchange the dime for two nickels.

So far you have two sets of coins worth 15¢. Work your way through the nickels by exchanging one of them for five pennies.

You now have three sets of coins. Take another nickel and exchange it for another five pennies.

That makes four sets of coins. Now take the last nickel and exchange it for five pennies.

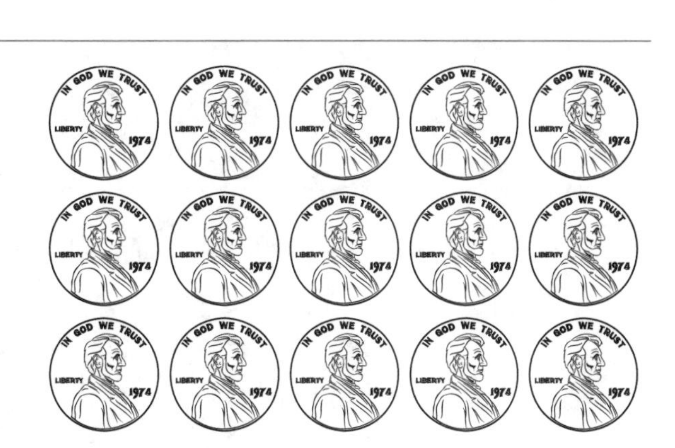

You have 15 pennies and that's as far as you can go. You have made five equivalent sets of coins worth 15¢. Good work!

Let's look at them altogether.

Suppose you had 20¢ in your pocket. How many different sets of coins could you have?

The largest-value coins used to make 20¢ are two dimes.

Then take one of those dimes and exchange it for two nickels

So far, we have two sets of coins. Next, take the other dime and exchange it for two nickels to make four nickels.

Now work your way down with the nickels. Take one nickel and exchange it for five pennies.

Then take another nickel and exchange it for five pennies.

You have 10 pennies and two nickels.

Exchange another nickel for five pennies.

You now have 15 pennies and one nickel.

Take the last nickel and exchange it for five pennies.

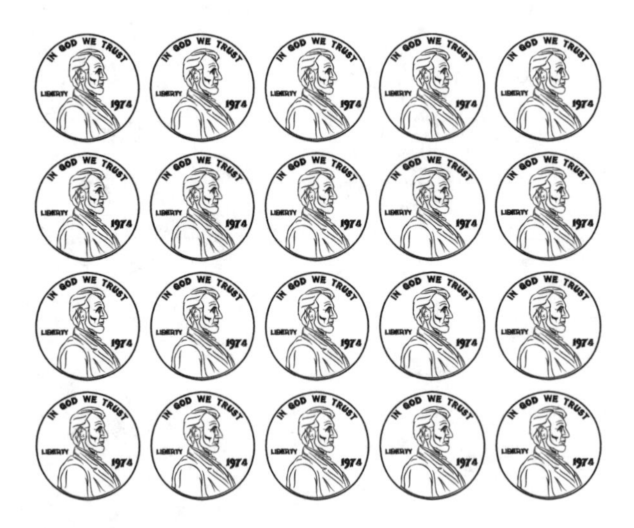

You have 20 pennies and have finished your task. How many sets of coins did you find?

That's seven equivalent sets of coins!

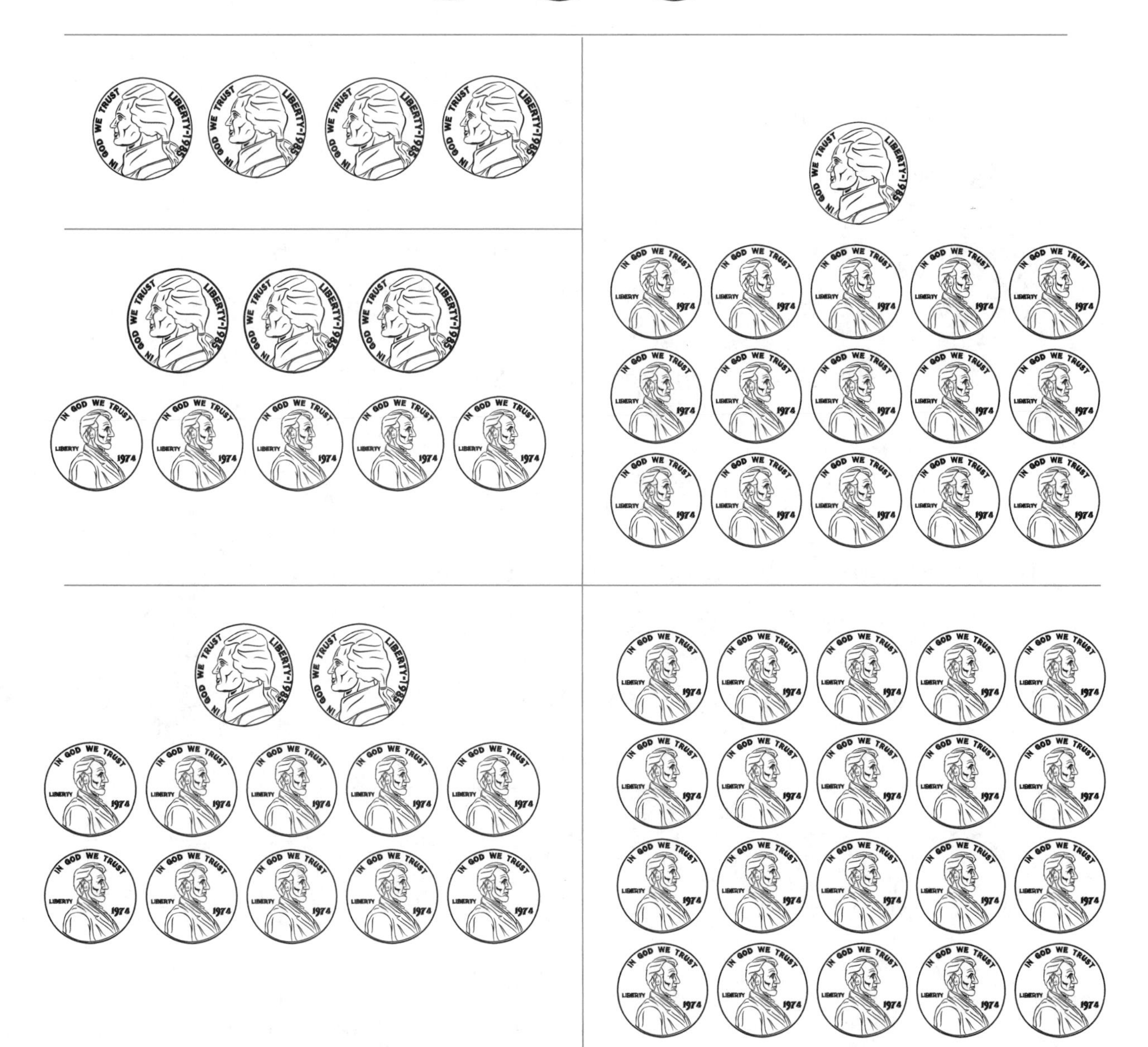

Let's Try It!

Set #1

Write three equivalent sets of coins for each of the following amounts.

1. 18¢ ____________ / ____________ / ____________
2. 25¢ ____________ / ____________ / ____________
3. 32¢ ____________ / ____________ / ____________
4. 21¢ ____________ / ____________ / ____________
5. 14¢ ____________ / ____________ / ____________

Think About It!

What do the equivalent sets of coins for the following amounts all have in common? 21¢, 26¢, 33¢, 38¢

Answers are on pages 179–180.

ORDER AND COMPARE SETS

Now that you are comfortable with equivalent sets of coins, we can look at ordering and comparing them.

Look at the two sets of coins shown below.

#1

#2

How much money is shown in box #1?

You were right if you said 31¢.

How much money is shown in box #2?

You were right if you said 27¢.

You remember from Chapter 2 the signs for greater than and less than. Use one of them now to draw between the boxes to show a correct math statement.

Since 31¢ is greater than 27¢, you should have used the greater than sign that looks like this. >

Let's look at some more sets of coins.

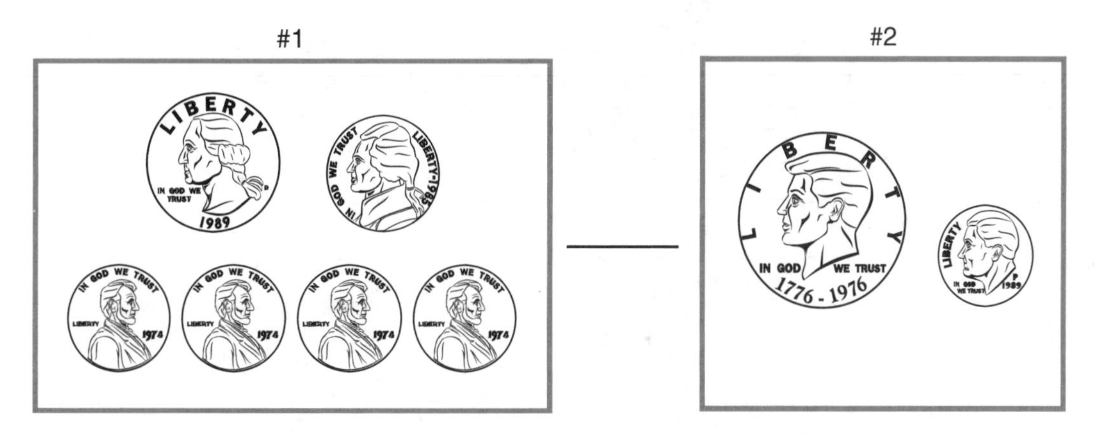

First, figure out how much money is in each box. Then draw the symbol between them to make a true math statement.

How did you do?

Did you find 34¢ in box #1 and 60¢ in box #2? If you did and drew the less than symbol between them, you were right!

Let's Try It! Set #2

Write <, >, or = on the line provided to compare the amounts in these pairs of boxes.

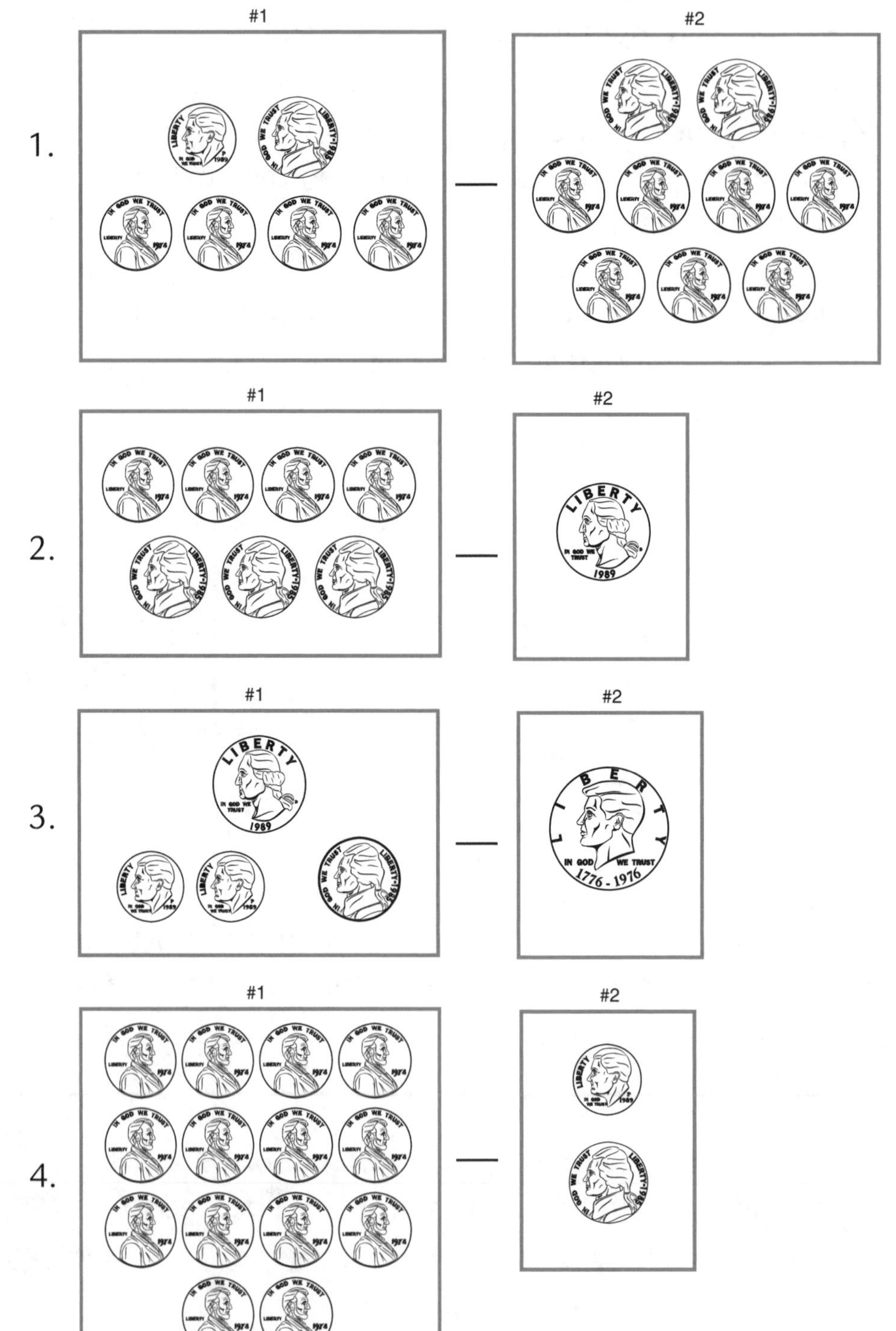

5.

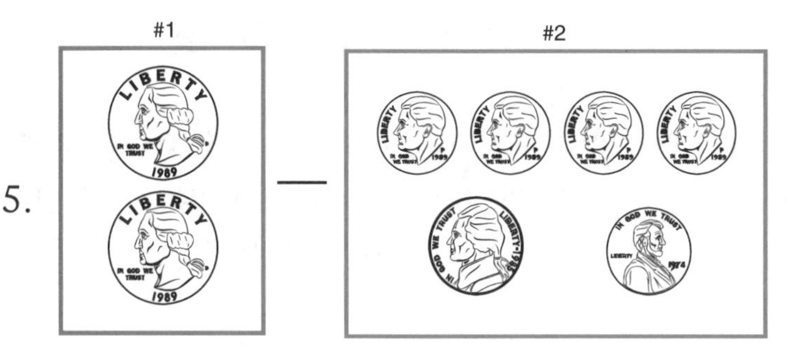

Think About It!

Does the number of coins in a box decide the value?

Answers are on page 180.

READING MONEY

Do you remember what we learned about reading numbers in Chapter 2? We learned that you never say the word *and* when reading numbers aloud except in special cases. Here is one special case! When you have an amount of money that contains bills as well as cents, you say the word *and* between the two to separate them. Here is an example: $5.95. To read this correctly, you would say five dollars and ninety-five cents.

It is usually easier to compare and order money amounts when dollar bills of any sort are added to your money. For instance, $5.23 is greater than $1.98. The dollar amount is what counts.

MAKE CHANGE

When you were younger, did you ever pretend to run a store? If so, you had to make change when your customers made their purchases. Now that you are older, you have to know how to make change with real money.

Let's look at two ways to make change. The first is one you are most likely to use as a student in school with paper and pencil in front of you. Here's how it works.

Nell bought some gum for 65¢. She paid for it with a one-dollar bill. How much change did Nell get?

You can find out the amount of change by subtracting 65¢ from $1. It will look like this.

$$\begin{array}{rrr} & \overset{9}{\cancel{10}} & 10 \\ \$\cancel{1}. & \cancel{0} & \cancel{0} \\ -\ . & 6 & 5 \\ \hline \$\ . & 3 & 5 \end{array}$$

Nell would get 35¢ change.

Sometimes larger amounts are involved as in this problem:

Caleb bought a game for $3.39 and paid with $5. How much change did Caleb get?

The answer on paper would look like this.

$$\begin{array}{rrr} 4 & \overset{9}{\cancel{10}} & 10 \\ \$\cancel{5}. & \cancel{0} & \cancel{0} \\ -3. & 3 & 9 \\ \hline \$1. & 6 & 1 \end{array}$$

Caleb would receive $1.61 change.

Most of the time in your life you will use the other common method of making change called **counting on**. Here's how this works.

Let's go back to Nell and her gum. The clerk could give her the change back by counting on like our friend Pethagerus. Begin with the amount and count on with Pethagerus to get to $1.

One dime and one quarter equal $.35, so Nell will get $.35 change.

How about Caleb's change? He spent $3.39. Count on from that amount like this:

$5.00

One penny, one dime, two quarters, and one dollar equal $1.61, so Caleb will get $1.61 in change.

Let's Try It!

Set #3

Use the counting on method to find the change for each problem.

1. Julia bought some marbles for 43¢ and paid with $1. How much change did Julia get? ________
2. Liam bought some cards for $3.79 and paid with $5. How much change did Liam get? ________
3. Sara bought a t-shirt for $8.58 and paid with $10. How much change did Sara get? ________
4. Dana bought a slice of pizza for $1.75 and paid with $5. How much change did Dana get? ________
5. Abby bought her lunch for $6.55 and paid with $10. How much change did Abby get? ________

Think About It!

What is the least number of coins you need to make 45¢?

Answers are on pages 180–181.

MENTAL MONEY MATH

As you get older and have more money of your own to use every day, you will do a lot of counting in your head or mental math. It is an easy thing to practice at home. Most of us have a jar somewhere in our house for spare coins.

Take a handful of coins without looking and spread them out in front of you. Arrange them in order from the greatest to the least value and begin to count. Look at this example.

Suppose you pulled out the following coins. Arrange them as shown.

Count the quarters first to get 50¢. Then add the dime to get 60¢. Add the three pennies to make 63¢.

If you count coins from greatest to least in value, you will have to memorize counting in quarters first. Then you will be counting mostly in fives and tens for nickels and dimes. The pennies are easy to add by counting on in ones.

Let's Try It!

Set #4

Count these coins in your head by touching the coins as you go.

1. ________

2. ________

3. ________

Think About It!

How is it possible to have the same answer twice with different sets of coins?

Answers are on page 181.

Chapter 7
HOW DO YOU MEASURE UP?
A Chapter About Measurement

Terms and Definitions

Customary system: The measurement system used in the United States.

Linear measurement: Measurement in length.

Inch: The smallest unit of length in the customary system.

Foot: A unit of length in the customary system: 12 inches = 1 foot.

Yard: A unit of length in the customary system: 3 feet = 1 yard.

Mile: A unit of length in the customary system: 5,280 feet = 1 mile.

Cup: A small unit of measure of capacity in the customary system.

Pint: A unit of measure of capacity in the customary system: 2 cups = 1 pint.

Quart: A unit of measure of capacity in the customary system: 2 pints = 1 quart.

Gallon: A unit of measure of capacity in the customary system: 4 quarts = 1 gallon.

Ounce: The smallest unit of measure of weight in the customary system.

Pound: A unit of measure of weight in the customary system: 16 ounces = 1 pound.

Metric system: The measurement system used in almost every country in the world.

Meter: The standard linear measure of the metric system.

Centimeter: A small unit of linear measure in the metric system: 100 cm = 1 meter.

Millimeter: A small unit of measure in the metric system: 1,000 mm = 1 meter.

Decimeter: A unit of linear measure in the metric system: 10 dm = 1 meter.

Kilometer: A unit of linear measure in the metric system: 1,000 m = 1 km.

Liter: A unit of capacity in the metric system.

Milliliter: A small unit of metric capacity: 1,000 mL = 1 L.

Gram: A small unit of weight or mass in the metric system.

Kilogram: A measure of weight or mass in the metric system: 1,000 kg = 1 g.

Temperature: Measurement of how hot or cold something is.

Thermometer: The instrument used to measure temperature.

Degrees Celsius: The metric unit of measuring temperature.

Degrees Fahrenheit: The customary unit of measuring temperature.

WHAT IS CUSTOMARY MEASUREMENT?

Your parents surprise you with a Saturday project. They tell you that if you accurately measure your room, you will be given new bedroom furniture to replace the stuff with ducks on it that you have now. This is a serious offer that you do not want to miss! What tools do you need to get this right? In this country, you need a yardstick and a ruler to measure the length and width of your room. While you're at it, measure the height of your walls so you can get something tall if you want.

Measuring Distance

Linear measurement is what you are doing when you measure how long something is. The United States is one of the few countries in the world that use inches, feet, yards, and miles. We'll talk more about that later.

Here's how it works: The smallest length of measurement in our **customary system** is called the **inch**.

0 1 2 3 4 5 6 7 8 9 10 11 12

It takes 12 inches to make 1 **foot** and 3 feet to make 1 **yard**.

0 1 2 3 4 5 6 7 8 9 10 11 1 13 14 15 16 17 18 19 20 21 22 23 2 25 26 27 28 29 30 31 32 33 34 35 3

The longest length of measurement is the **mile**. There are 5,280 feet in 1 mile.

0 — 1 mile — 5,280 feet

This is all you really need to know.

Which is the best unit to use to measure the length of your pencil?

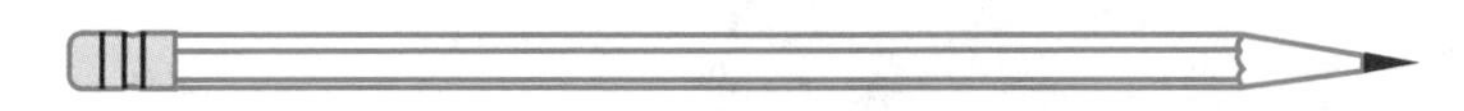

If you said the inch you were right.

Which is the best unit to use to measure the distance between your house and your friend's home on the other side of your town?

If you said the mile you were right.

Which unit is best to measure the length of the desk in your school classroom?

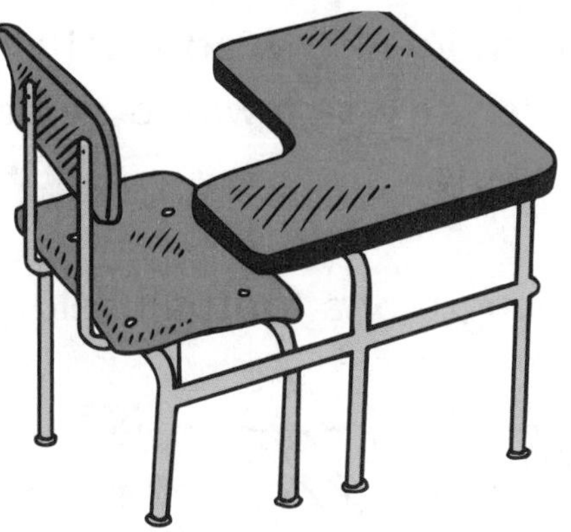

If you said the foot you were right.

Which is the best unit to use to measure the length of the hall in your school?

If you said the yard you were right.

Let's Try It!

Set #1

Using what you know about the customary units of length, decide which would be best to use for measuring the following:

1. The length of a football field ___________

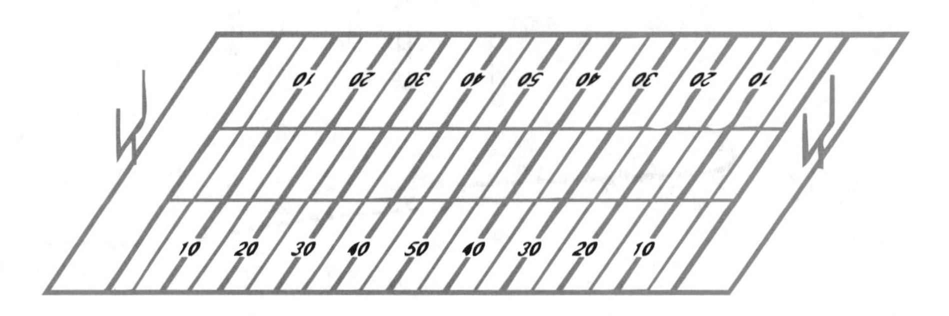

2. The length of a crayon ___________

3. The length of a telephone cord in your kitchen ___________

4. The distance between your town and the town next to yours ____________

5. The length of your math textbook ____________

Think About It!

Why is it necessary to measure in fractions of inches but not fractions of feet or yards?

Answers are on page 181.

Measuring Capacity

The customary system of capacity is loaded with different units. It is not easy to keep them all straight. The smallest unit is the **cup**.

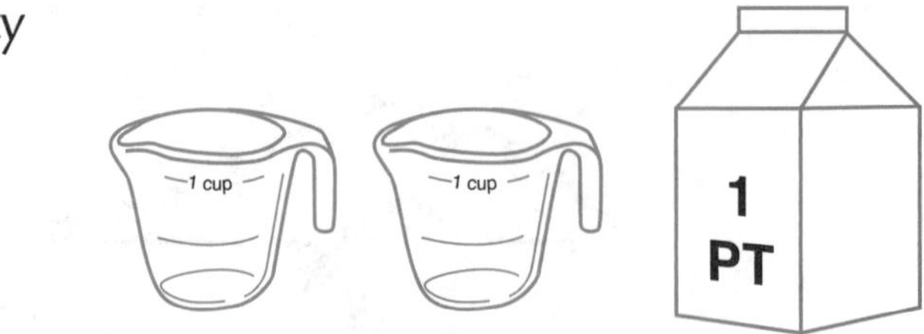

It takes two cups to make a **pint**.
It also takes two pints to make a **quart**.

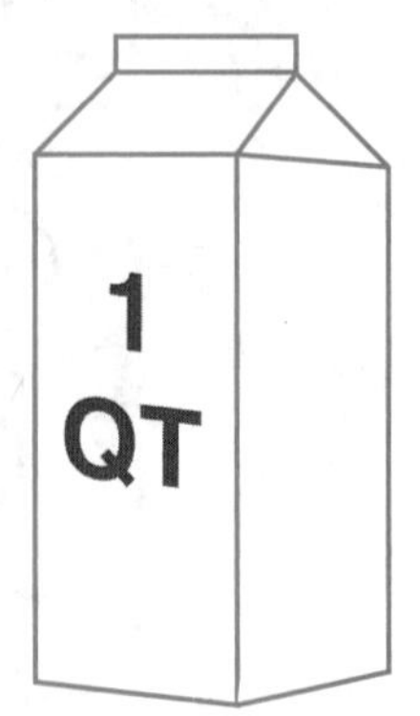

Now comes the easy part.
It takes four quarts to make a **gallon**.

Do you notice the beginning of the word *quarter* in quart? Did you remember that it takes four quarters to make a whole and four quarters to make a dollar? That should help!

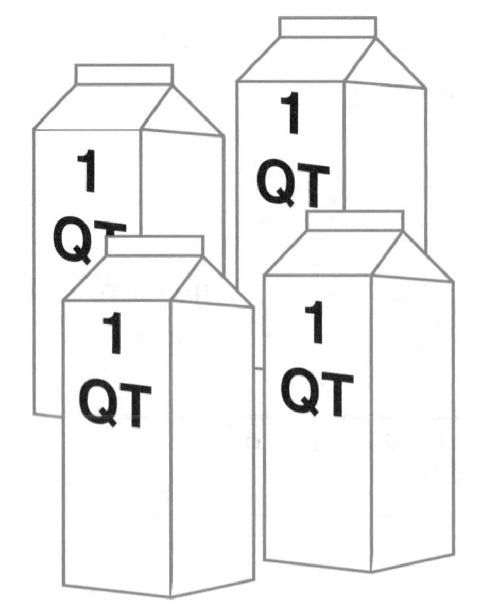

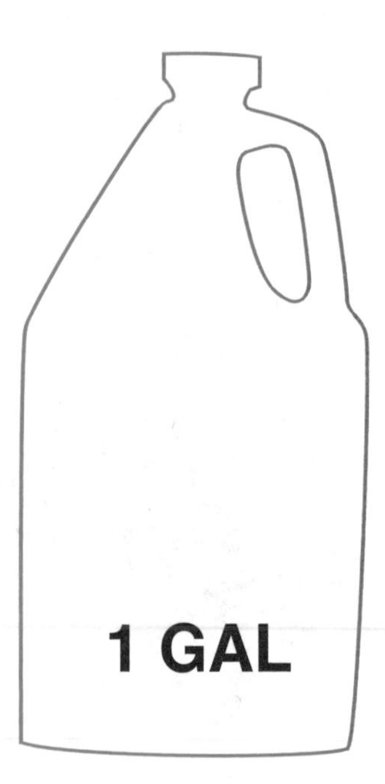

Measuring Weight

The customary system of weight is less burdened with units than the customary system of capacity. We will use only **ounces** and **pounds**. It takes 16 ounces to make 1 pound. If that isn't strange enough, the abbreviation for ounces is oz and the abbreviation for pounds is lb.

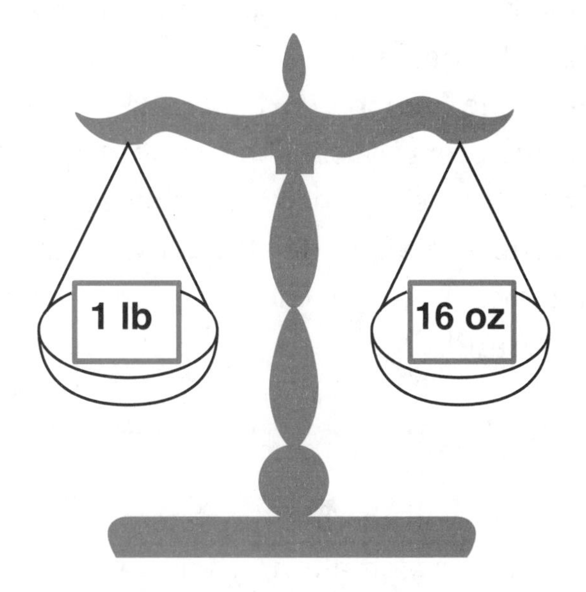

What Is Metric Measurement?

Almost every country in the world other than the United States uses the **metric system**. It was devised and first used in the late 1700s in France. All the measurements are based on powers of ten just like our number system. Once you learn a few prefixes and what they mean, you can do anything you want with this system. The best part is that you don't have to do any multiplying. The digits remain the same, and the decimal point changes to change the unit of measure. How easy is this?

Here Are the Prefixes You Need

Let's start with the prefixes you need to know. Once you've learned them, you can use them in all three kinds of measurement. The first one to learn is *centi*. You can remember this by thinking of a dollar. How many *cent*s are in a dollar? If you said 100 you were right. The prefix *centi* means one hundredth of. The next important prefix is *milli*. For many of us *milli* is the beginning of the word *mill*ion, but be careful here. This comes from the French word *mille*, which means thousand, so the prefix *milli* means one thousandth of. Another prefix to use is *deci*. Think of the first three letters here and then answer this question. How many years are there in a *dec*ade? If you said ten you were right. The prefix *deci* means one tenth of. The last prefix you need is *kilo*. This prefix means one thousand times.

Measuring Distance

Linear measurement is easy in the metric system partly because it is so flexible. You can measure very small things and very long distances. The **meter** was the original unit of measure. Look at the table below to see the metric prefixes at work for you.

1 meter
10 centimeters
0 1 2 3 4 5 6 7 8 9 10
decimeters

1 meter (m)	1000 millimeters (mm)
1 meter (m)	100 centimeters (cm)
1 meter (m)	10 decimeters (dm)
1 centimeter (cm)	10 millimeters (mm)
1 decimeter (dm)	10 centimeters (cm)
1 kilometer (km)	1000 meters (m)

1 km

Let's Try It!

Set #2

Choose the best unit to measure these things.

a. millimeters
b. centimeters
c. decimeters
d. meter
e. kilometers

_______ 1. The diamond in a ring.

_______ 2. The height of your school building.

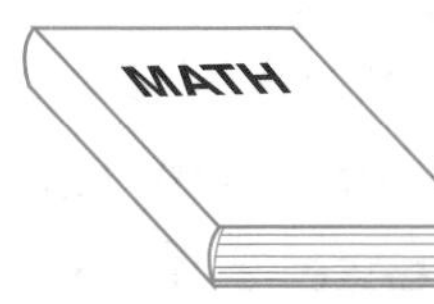

_______ 3. The width of your math textbook.

_______ 4. The length of your pencil.

_______ 5. The distance of the bus ride you take on a school field trip.

Think About It!

How do you explain to someone who doesn't know the system how to decide which unit to use?

Answers are on page 181.

MEASURING CAPACITY

Capacity or liquid measurement is easy in the metric system, too. As a matter of fact, most liquid containers have metric measurements listed next to the customary. The basic unit of capacity is called a **liter**. Soda bottling companies have begun to sell their drinks in liter bottles. Most of the time all you need to know is this:

1 liter (L)	1000 milliliters (mL)

"Wow! A milliliter is really small!"

Did you notice that a capital L is used to stand for liter? Can you think why it is used rather than a lowercase l? If you said that the lowercase l could look like a 1 (one) you were right.

MEASURING WEIGHT

Weight or mass measured in the metric system is also simple. The basic unit is called the **gram**. A gram weighs about as much as a regular paper clip.

Here's all you need to know for metric mass or weight:

1 kilogram (kg)	1000 grams (g)

Let's Try It!

Set #3

Choose the best unit of measurement for the following:

1. A large pitcher of milk. ____
   a. milliliter
   b. liter

2. A balloon. ____
   a. gram
   b. kilogram

3. A teaspoon. ____
   a. milliliter
   b. liter

4. A whole watermelon. ____
   a. gram
   b. kilogram

5. A chalkboard eraser. ____
   a. gram
   b. kilogram

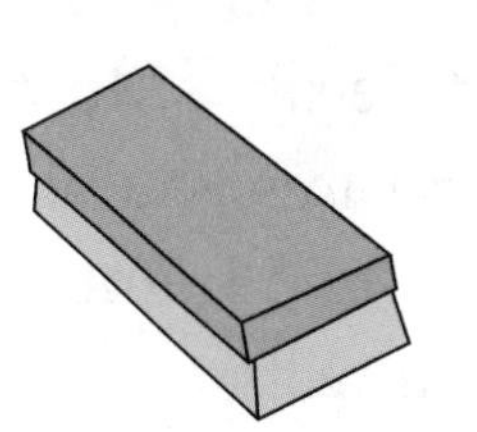

Think About It!

A box of cereal has been marked with the weight of 35 kg. Does this make sense? Explain why or why not.

Answers are on pages 181–182.

TEMPERATURE

There is one aspect of measurement that is magical. Almost every day we measure something that we can't really pick up and touch the same way we pick up a baseball or a pencil. Can you think of it?

Temperature! When you read a **thermometer** you are measuring how warm or cold the air is. Of course, you can measure the temperature of a liquid, a roast that is baking, or your body.

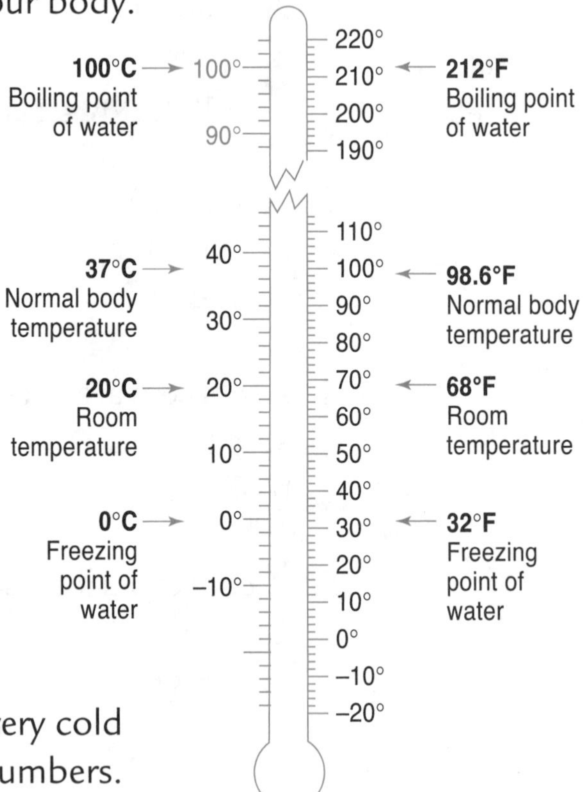

A thermometer has been drawn for you with **degrees Celsius** on the left and **degrees Fahrenheit** on the right. Some common temperatures have been marked on the thermometer.

Once again the metric system of measurement in degrees Celsius is the easiest to use. Just remember that water freezes at 0° Celsius and boils at 100° Celsius. Most of the other temperatures you will use fall between those two numbers. On a very cold winter day you will get to negative numbers. How cool is that?

Careful!

Don't be fooled by the 100° Celsius boiling temperature. A hot day at the beach will still be only about 30° Celsius.

An interesting fact is that your body temperature is 37° Celsius.

Chapter 8
HOW DO YOU
SHAPE UP?
A Chapter About Geometry

Terms and Definitions

Point: A location in space.

Line: A straight path of points.

Line Segment: A line with two endpoints.

Intersecting lines: Lines that cross each other.

Perpendicular lines: Lines that cross each other forming right angles.

Parallel lines: Two or more lines that travel the same distance apart.

Polygon: All closed figures (geometric shapes) with straight lines for sides.

Quadrilateral: Any polygon with four sides.

Square: A quadrilateral with four equal sides and four right angles.

Rectangle: A quadrilateral with equal opposite sides and four right angles.

Trapezoid: A quadrilateral with two and only two parallel sides.

Rhombus: A quadrilateral with four equal sides.

Parallelogram: A quadrilateral with equal opposite sides.

Pentagon: A polygon with five sides.

Hexagon: A polygon with six sides.

Octagon: A polygon with eight sides.

Face: A flat surface of a solid shape.

Rectangular prism: A solid shape with six rectangular faces.

Cube: A rectangular prism with six square faces.

Cylinder: A solid shape with two congruent circular faces and a curved side connecting them.

Cone: A solid pointed shape with one circular face.

Pyramid: A solid shape with triangular faces and either a square or triangular base.

Sphere: A solid shape that is rounded like a ball.

Symmetry: Something that occurs when one half of a figure is the mirror image of the other half.

Line of Symmetry: The line that divides a symmetrical figure in half.

Congruent: Two figures have the same size and shape.

Similar: Two figures having the same shape and the same or different size.

Endpoints: Points that mark the ends of a line segment.

Triangle: A polygon with three sides.

THE MAGIC OF GEOMETRY

Did you know that geometry is magic? Did you ever think about that? The magic of geometry is what you don't see. You have to use your imagination. It starts at the very beginning.

Important!
All letters used for labels in geometry are capital letters.

GEOMETRY BEGINS WITH THE POINT

A **point** is a location in space and is labeled with a letter like this.

• A

A point is magical because even though it is a location in space, it can't always be seen.

Even more magical than that is the fact that no matter how close two points are to each other, you can always place another point between them!

LINES AND SEGMENTS

When you place a lot of points together you have a **line**. You can draw a line like this.

The magical thing about a line is that it goes on forever in both directions. That is why you will always see the arrows on both ends. As you hold this book and look at the line drawn above, imagine that it travels on forever. It goes through the wall and out of your house. It travels across your lawn and through the fence.

It never ends!

There are times when we need only a small part of a line. We call that a **line segment**. We draw it like this.

A B

Do you see the dots on each end? Those are called the **endpoints**. As you can see, they are also labeled with letters.

Have you thought about how many times you see line segments every day? Look at the floor. Are there wooden planks or square tiles? They have line segments running around them.

PARALLEL, PERPENDICULAR, OR NOT?

Whenever these lines cross each other they are called **intersecting lines**. You can draw them like this:

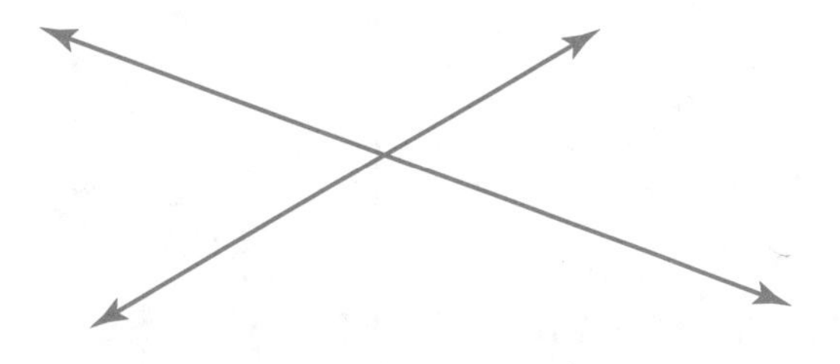

Look at this map. The place where two streets cross is called an intersection.

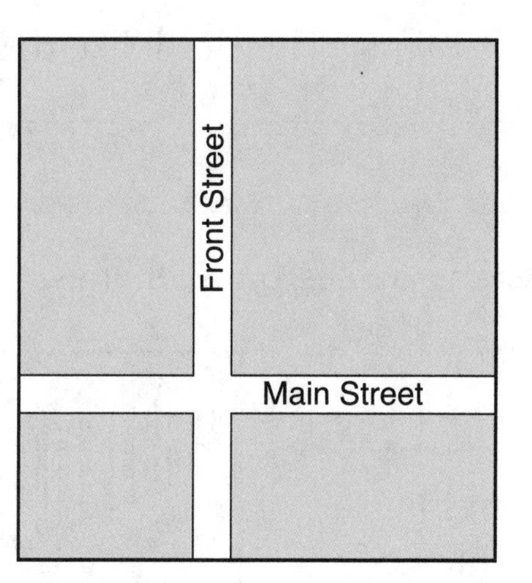

Other times you see lines that cross each other and form square corners we call right angles. They are called **perpendicular lines** and they look like this.

Can you think of times you see perpendicular lines?

How about these?

Can you think of other perpendicular lines you see?

Sometimes you see more than one line traveling together at the same time. When these lines stay the same distance apart as they travel they are called **parallel lines**. They look like this:

Can you think of times when you see parallel lines?

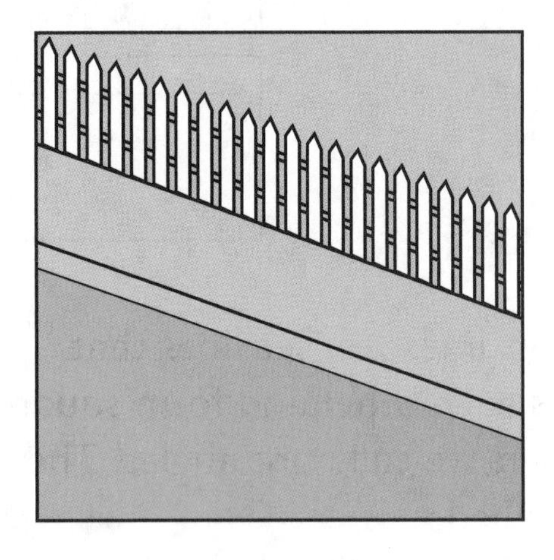

How about these?

Can you see the parallel lines in these drawings? Can you think of other sets of parallel lines?

Here is a handy tip! The word parallel has parallel lines in it. Do you see the two lowercase L's together in the word?

parallel

Let's Try It!

Set #1

Identify each of these drawings using the words point, line, line segment, intersecting lines, parallel lines, or perpendicular lines.

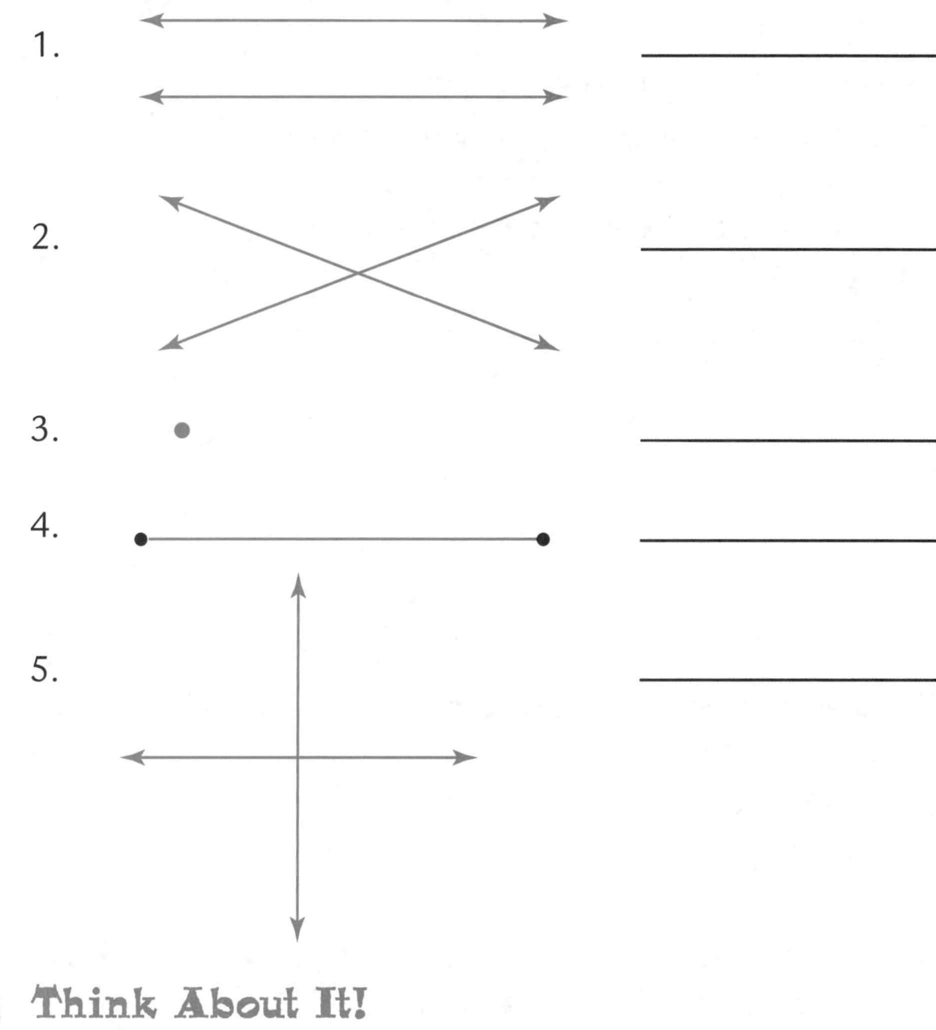

Think About It!

Are all intersecting lines perpendicular? Why or why not?

Answers are on page 182.

PLANE SHAPES

Ever since you were a very young person, you probably have played with geometric shapes. Along the way you learned some of the names. Now, as an older person, you will study these shapes a little more and add some new ones to your list.

Let's begin with how the shapes are formed. All plane geometric shapes with line segments for sides are called **polygons**.

The simplest of these shapes is the **triangle**. Look at the drawings of triangles.

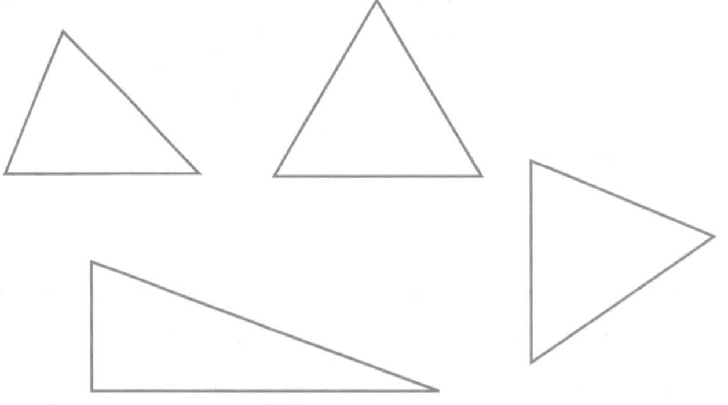

The next polygon you learned about was probably the **square**. This shape and all shapes with four sides are called **quadrilaterals**. A square has the same length on each of its four sides. Look at the drawings of squares as you may see them in real life.

A shape that is related to the square is a **rectangle**. It also has four sides, but usually two sides are longer than the other two sides. Look at the drawing of this rectangle.

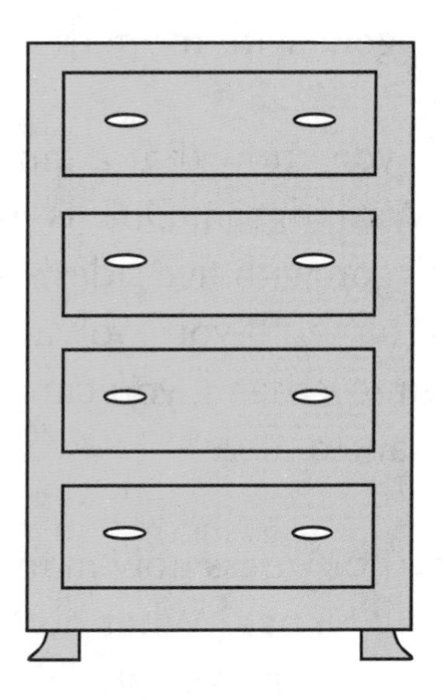

A quadrilateral that has a funny name is the **trapezoid**. Two of its sides are parallel, and the other two are not. If you have worked with pattern blocks, then you may remember that the red block is a trapezoid. Look at some drawings of trapezoids.

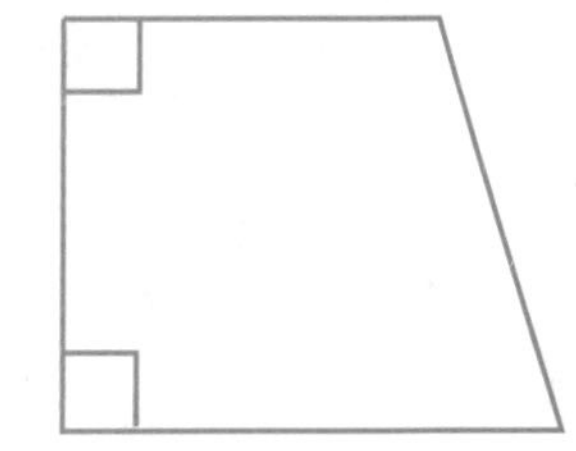

Have you ever seen a square that is leaning a little? That shape has a special name. It is called a **rhombus**. You may call it a diamond. Look at the drawings right.

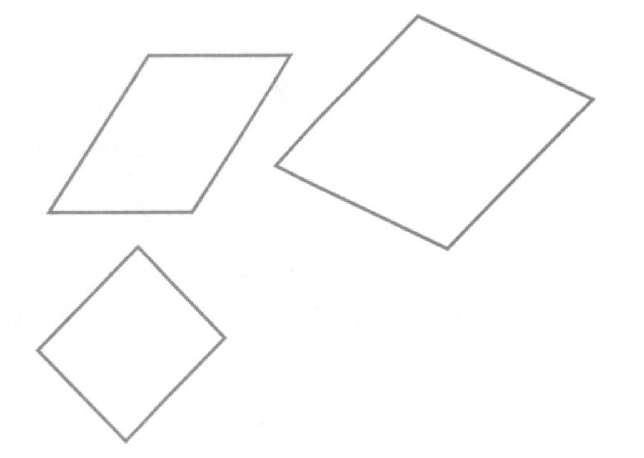

Did you know that there is a special word for a rectangle that leans a little? It is called a **parallelogram**. Look at the drawings below.

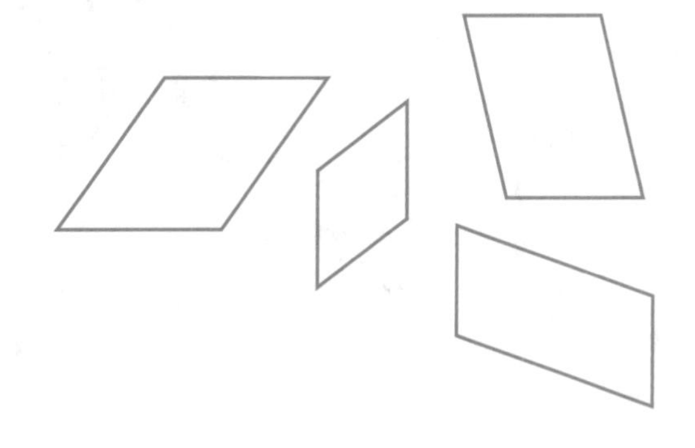

Polygons with more than four sides are very interesting.

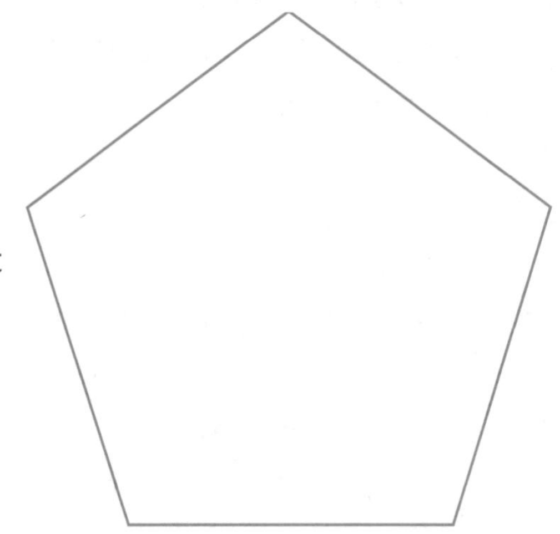

Did you know that a famous building in Washington, D.C. is named after a polygon with five sides? It is called a **pentagon**. If you look at the building from overhead, you can see the shape it is named after.

Can you guess how many sides a **hexagon** has? What number has the letter X in it? Yes, the hexagon has six sides! Look at the drawings below.

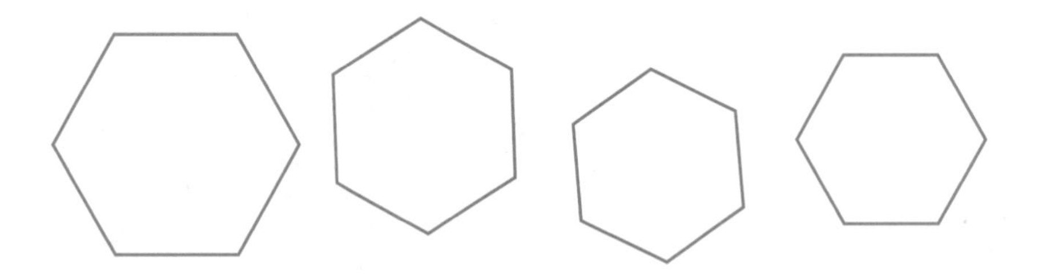

How many arms does an octopus have? Can you guess how many sides an **octagon** has? Yes, the octagon has eight sides. Do you know what sign uses the octagon for its shape? Look at the drawing below to find out.

Let's Try It!

Set #2

Identify each of these polygons by name.

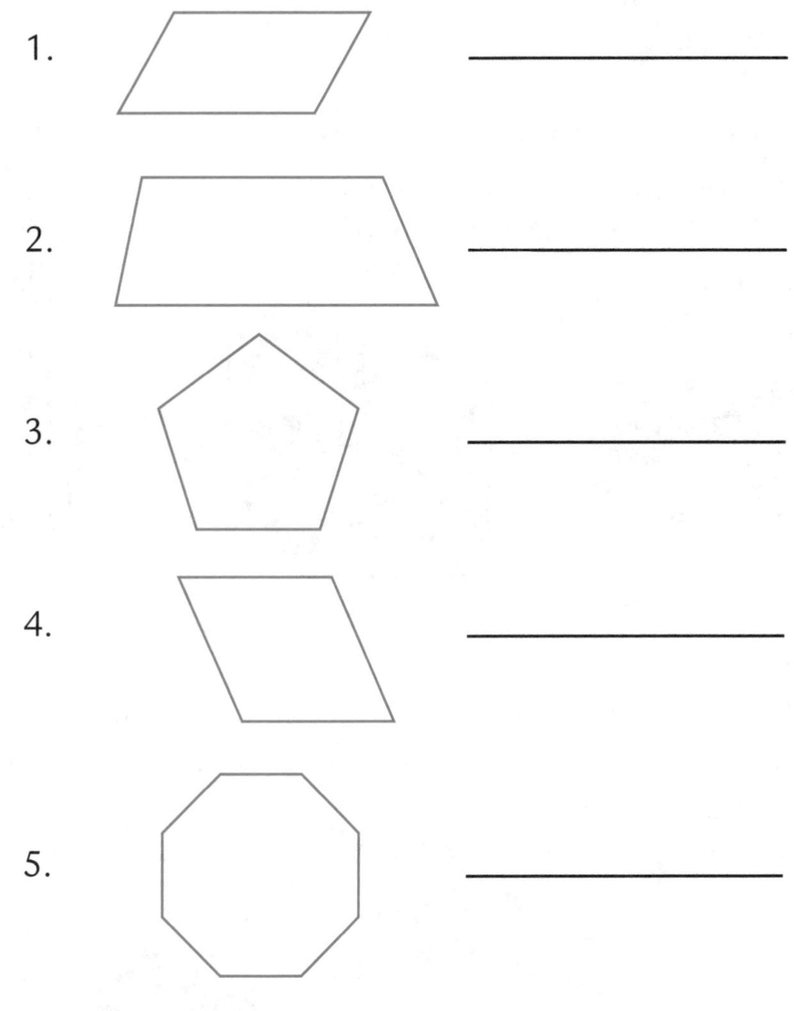

1. ________________

2. ________________

3. ________________

4. ________________

5. ________________

Think About It!

Is a rhombus a special kind of parallelogram? Explain why or why not.

Answers are on page 182.

SOLID SHAPES

When you were a baby, the first toys you played with were most likely solid geometric shapes. Wooden building blocks come in all sorts of geometric shapes.

The most common of these is the **rectangular prism**. Most presents come in boxes that are this shape. Look at the drawings below of rectangular prisms.

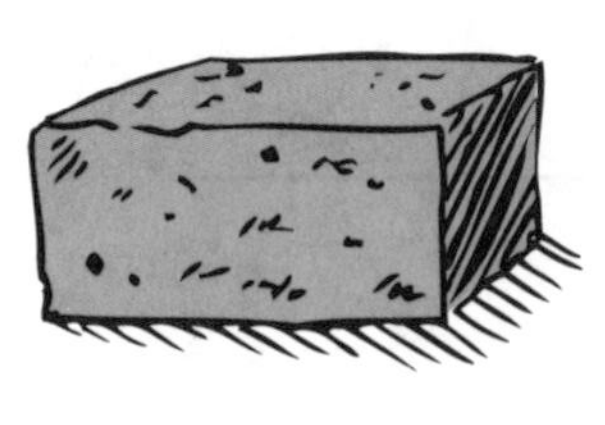

A block that is easy to build with is the **cube**. It is easy to build with because it is the same length on all sides. Look below at the drawings of some cubes.

If you go to your kitchen and open any cabinet, the chances are you will see a **cylinder** sitting on the shelf. It is a solid shape that has two flat sides called **faces** that are circular and rounded sides connecting them. Did you guess that this is the shape of a can of soup? Look at the drawings below.

It is a hot summer day and you would like to cool off with some ice cream. Can you imagine which shape we are going to talk about next? Yes, ice cream is often served in a shape called the **cone**. Look at the drawings below the see some more cones.

Many, many years ago some people made buildings of these shapes that modern people still travel many miles to visit. This shape is called the **pyramid**. Pyramids are also the shape of some tents and other things in daily life. Look at the drawings below.

A solid geometric shape that sometimes bounces is called the **sphere**. Yes, this is a shape you first knew as a ball. It does not work well as a building block because it rolls. Can you think of some things you know that come in this shape? Look at the drawings below.

Let's Try It!

Set #3

Identify each of these pictures as a solid geometric shape.

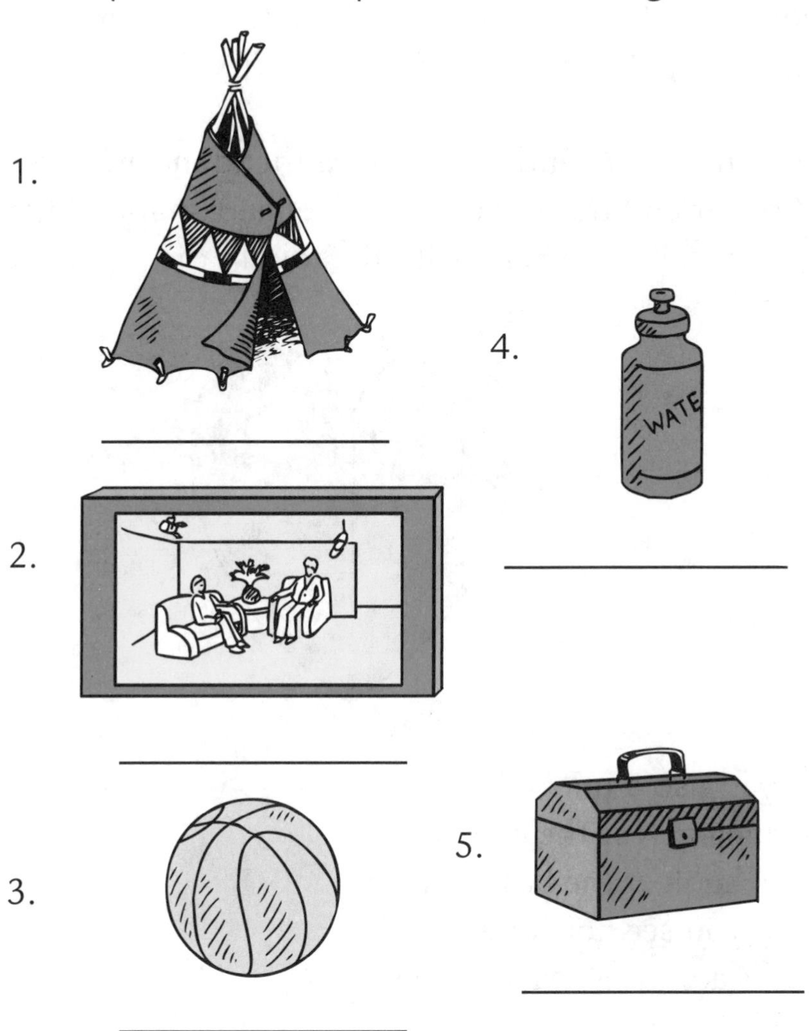

Think About It!

Compare the number of faces that a cube has and that a rectangular prism has. What do you notice? Why do you think this happens?

Answers are on page 182.

SYMMETRY

Symmetry is found in nature and in many things made by people. Symmetry is a balance so that two sides of a shape will match if they are folded along the **line of symmetry**.

When you make a Valentine card for a friend most often you will fold a piece of paper and trace and cut half a heart shape. When you unfold it you have a complete heart with the fold in the middle. Look at the drawing below.

Symmetry is also a mirror image of something across the line of symmetry. Look at the heart drawn above. Do you see how one side of the heart is a mirror image of the other?

Look at the square. How many lines of symmetry does it have? To be sure, you can trace this square onto another piece of paper in order to fold it.

Did you find four lines of symmetry? Look at this drawing of the same square with the lines of symmetry drawn on it.

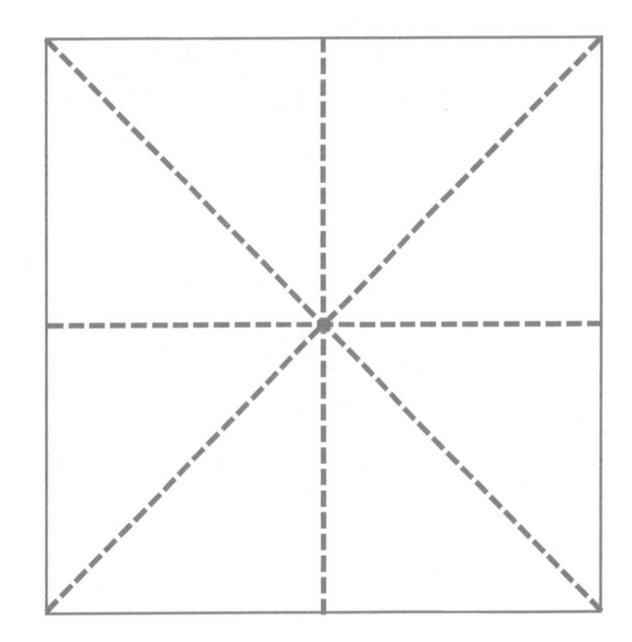

Now look at the rectangle drawn below. How many lines of symmetry do you think it has? Try the tracing method again so you are sure of your answer

Were you surprised by your answer? What happened when you tried to fold the rectangle on a diagonal line? Why do you think this happens? Look at the drawing below with the lines of symmetry drawn on them.

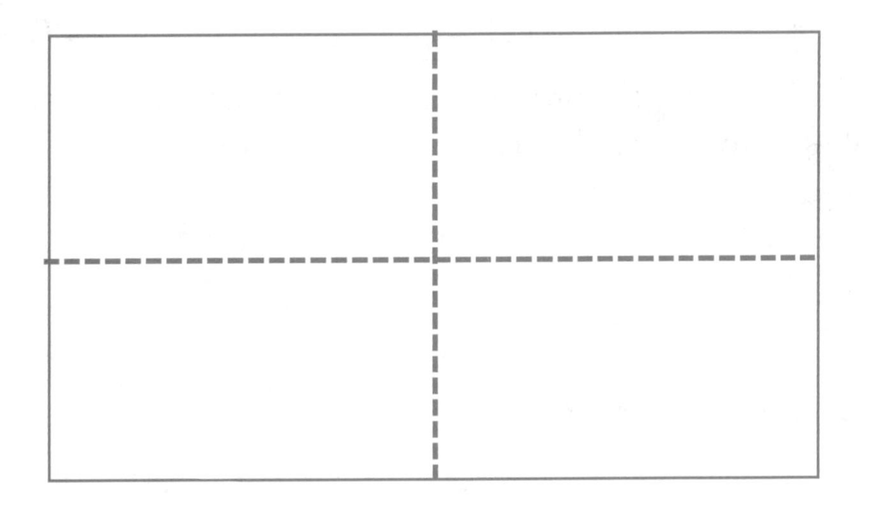

How about a parallelogram? How many lines of symmetry do you think it has? Look at the one that has been drawn for you, and trace it to be sure.

Did you find any lines of symmetry at all?
Why do you think this happens?

Let's Try It! Set #4

Count the number of lines of symmetry on each of the following shapes.

1.

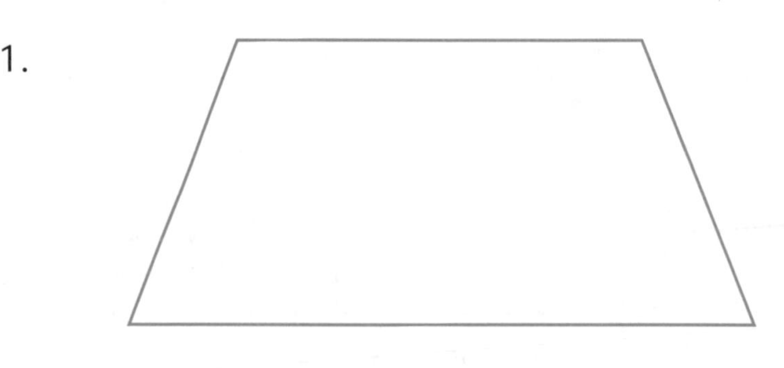

____

2.

3.

4.

5.

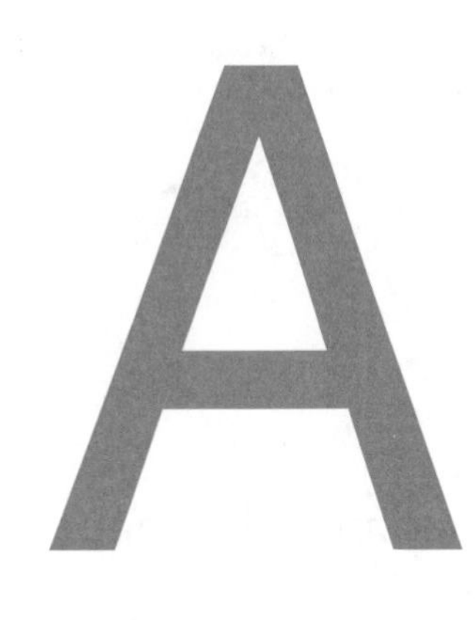

Think About It!

How many lines of symmetry does a circle have? Explain your answer. Hint: You may want to trace the circle to help you get your answer.

Answers are on page 182.

CONGRUENCE AND SIMILARITY

Have you ever seen a pair of twins that looked so much alike that you really couldn't tell them apart? That is nature's way of coming as close to the next topic as people can. When two shapes are the same size and shape we say they are **congruent**. Look at the drawings below.

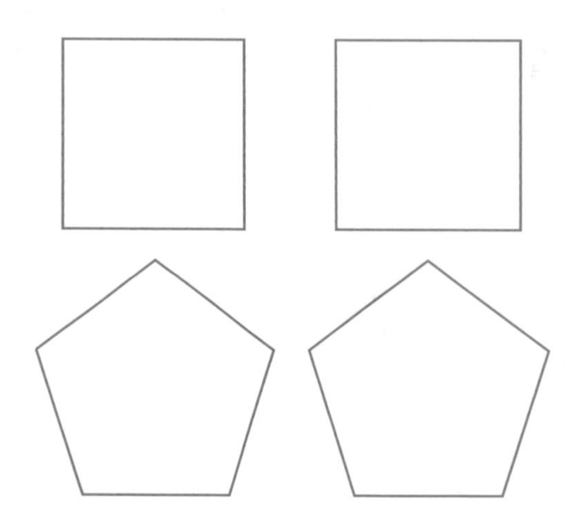

Sometimes two shapes are congruent, but they are placed on the paper in slightly different ways. It is a good idea to turn the paper or book to make sure. Our eyes are playing tricks on us. Look at the drawings below. Each pair of figures is congruent, but each shape is placed differently on the page. Turn the book to help you see better.

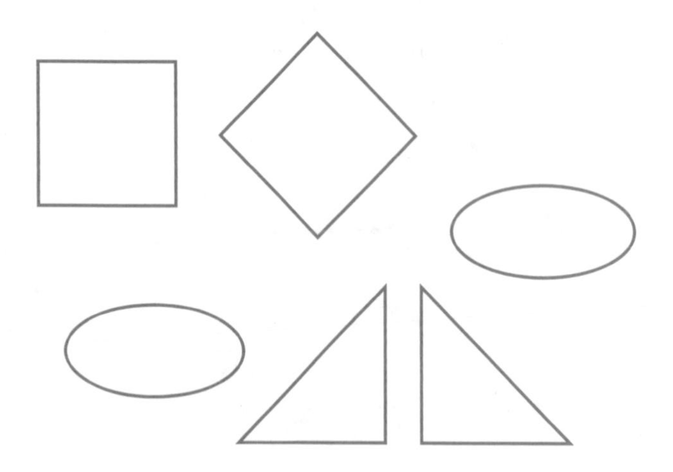

Look at the two circles drawn below. They are both circles, but one is larger than the other. They are called similar because they are the same shape but not the same size.

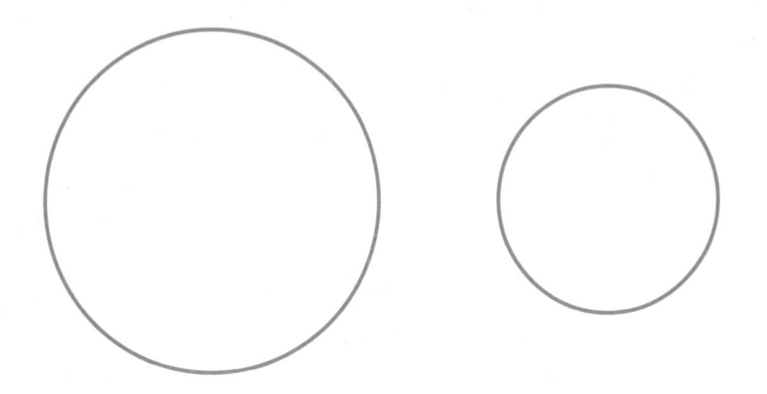

Now look at the figures drawn below. These are also similar.

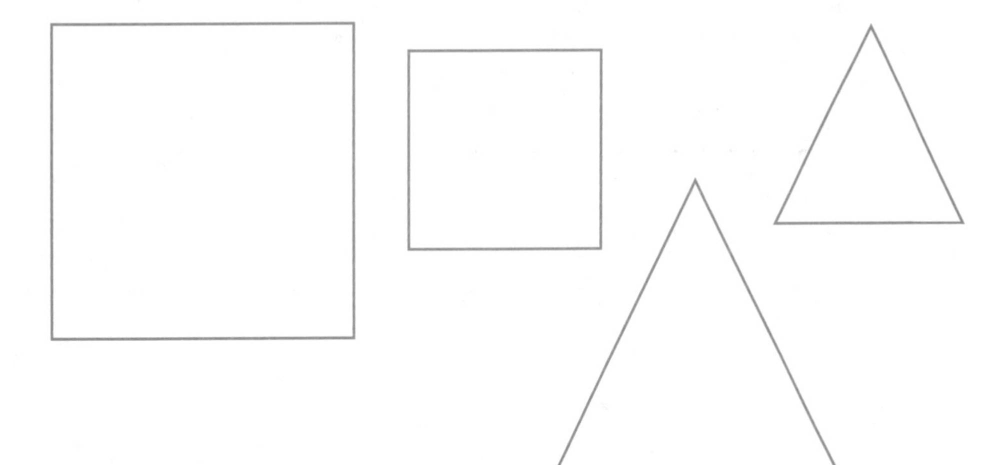

Careful!

Two shapes that are congruent are also similar!

Let's Try It!

Set #5

Decide if the following figures are congruent, similar, or neither congruent nor similar.

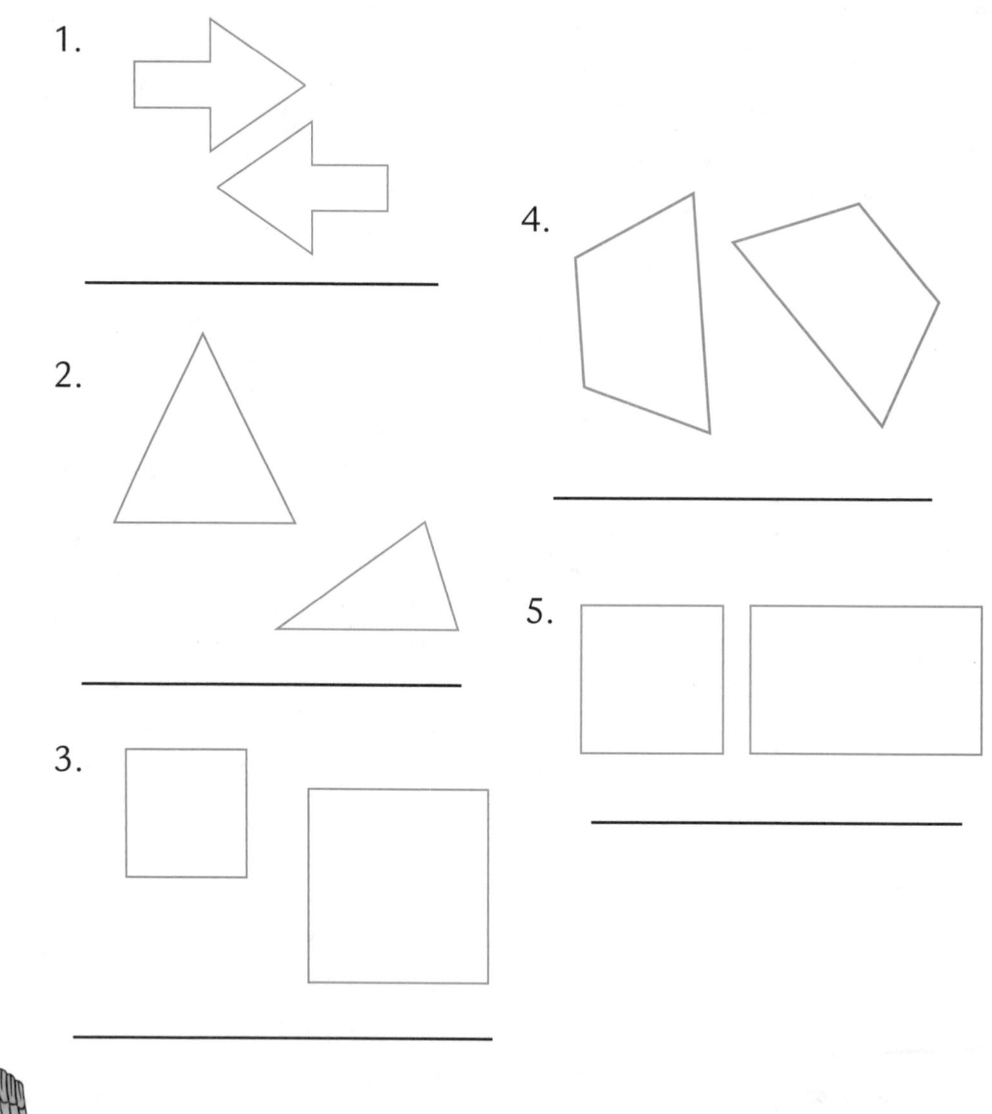

Think About It!

Are all circles similar? Why or why not?

Answers are on page 183.

Chapter 9

WHAT'S YOUR FAVORITE?

A Chapter About Graphs

Terms and Definitions

Data: Information collected about people or things.

Tally Chart: A chart that uses tally marks to record data.

Bar Graph: A graph that compares data using horizontal or vertical bars.

Interval: The difference between the numbers next to each other on a graph.

Pictograph: A graph that compares data using pictures.

Key: The part of a pictograph that explains the symbols.

Line Graph: A graph that uses a line to show a change in data over a period of time.

WHAT IS DATA?

What is your favorite ice cream flavor? What is your favorite television show? What is your favorite movie?

You may find yourself asking a question like this as you gather **data**. Simply put, data is information that you collect by asking a question.

HOW SHOULD I COLLECT DATA?

Once you have decided on the question to ask, you can begin to collect some data. As you ask your question, you should have a paper in front of you so you can organize the data as you get it. The best way to do this is with a **tally chart**. Put your question at the top of the chart and the choices you are going to offer as a list. Mark a tally for each answer you get. Be sure to bundle them by making the fifth tally mark across the first four.

If you are asking this question of each of your classmates, you should have a list of their names and check them off as you speak with them. That way you will not ask anyone the same question more than once. Look at the sample shown below.

The question asked was: What is your favorite flavor of ice cream?

Flavor	**Tally**	**Total**
Vanilla	~~||||~~ |||	8
Chocolate	~~||||~~ ~~||||~~	10
Strawberry	|||	3
Butter Pecan	~~||||~~	5

HOW SHOULD I ORGANIZE MY DATA?

BAR GRAPHS

Now that you have some data to work with, you can organize it in the form of a graph. This data will work well in a **bar graph**. You have probably worked with bar graphs since you started coming to school. They are easy and fun to do.

Before you do anything else, you should plan the **interval** you are going to use. The interval is the difference between the numbers on the graph. Check the size of the graph paper to see how many lines you have. An interval of 1 is the best if it's possible. Be sure to label the numbers on one axis. In this case they are the numbers of votes. The other axis is for the names of the ice cream flavors. Make sure that you include the general label of Ice Cream Flavors along with the names of the four flavors. The bars can be different colors. If you don't have colors, you can make a different design on each bar. Once you have filled in the bars, make sure you have written a title at the top. A good title for this graph might be Favorite Ice Cream Flavors.

This is how your graph might look.

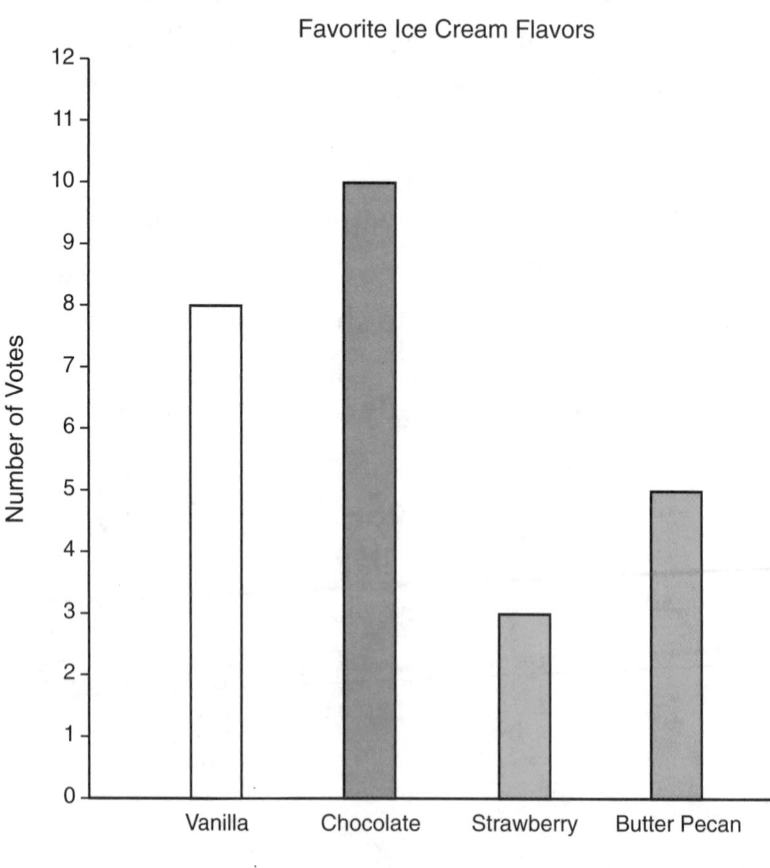

Let's Try It! Set #1

Use the data collected below to create your own bar graph. Don't forget to label the axes and give your graph a title.

The question asked was: What is your favorite season?

Season	Tally	Total
Winter	卌 III	8
Spring	III	3
Summer	卌 卌 II	12
Fall	IIII	4

Think About It!

Which two seasons when added together equal the total of a third season?

Answers are on pages 183–184.

Pictographs

Do you like to draw pictures? If you do, you are going to love making **pictographs**! Just like the name says, these are graphs that contain pictures. An important part of a pictograph is the **key**. This tells you the value of each picture in the graph.

Look at the data. Some students were asked to identify their favorite winter sport.

Sport	Tally	Total
Ice skating	~~||||~~ |||	8
Skiing	~~||||~~	5
Hockey	||||	4
Sledding	~~||||~~ ||||	9

It is important to choose a picture that is easy to draw because you have to draw it many times. It is also important to decide upon your key. For this graph the picture will be a snowflake. Each snowflake will be worth two votes, and a picture of half a snowflake will be worth one vote.

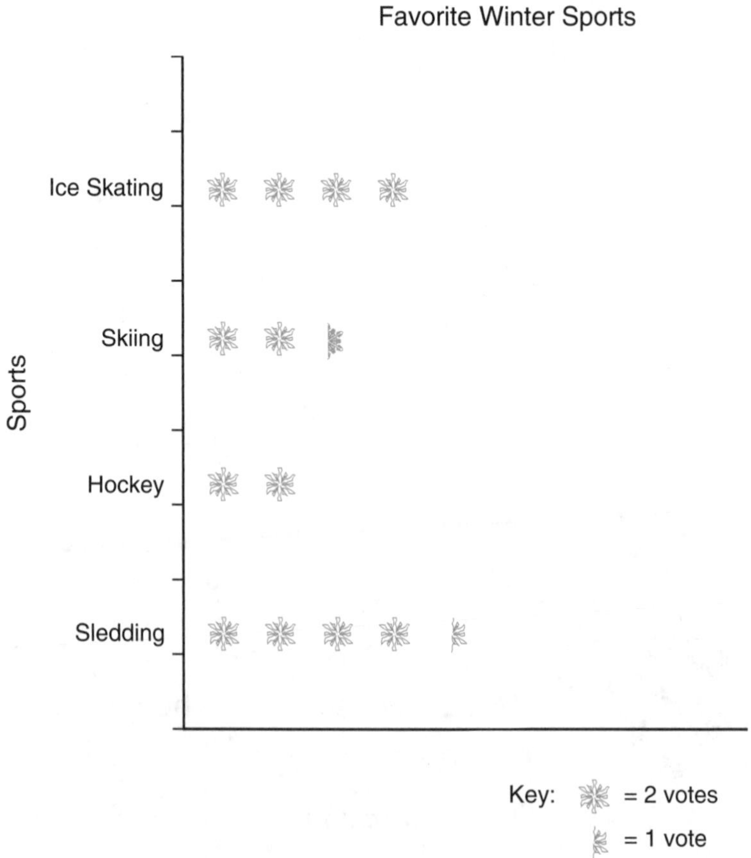

Careful!

Don't forget the title!

Favorite Winter Sports

Sports

Ice Skating

Skiing

Hockey

Sledding

Key: = 2 votes

= 1 vote

Let's Try It! Set #2

Some students were asked to identify their favorite summer sports. Use the data shown below to make a pictograph on the grid provided for you. Use a simple smiley face for your picture. It is shown in the key.

Sport	Tally	Total
Baseball	~~||||~~ |||	8
Swimming	~~||||~~ ~~||||~~	10
Biking	||||	4
Surfing	~~||||~~	5

Key: ☺ = 2 votes

◐ = 1 vote

Think About It!

Do you think that this data might be different if the students surveyed live in a state in the Midwest? Why or why not?

Answers are on page 184.

Line Graphs

Have you ever recorded the temperature outside your classroom? Did you notice changes from day to day? The best way to organize data like that is to make a **line graph**. A line graph shows data connected by a line. This is a very easy way to see changes in the data. A steep line shows a big change in either direction. A flat, or almost flat, line shows very little difference in change.

Mrs. Cosgrove's fourth grade class recorded the temperature outside its classroom at 1:00 P.M. every school day for a week. The results are shown in the table.

Day	**Temperature in Fahrenheit**
Monday	23°
Tuesday	28°
Wednesday	32°
Thursday	25°
Friday	20°

When setting up the grid to make this graph, you need to think about the interval again. This time the data is grouped from 20° to 32° Celsius. It's a good idea to use as much of the graph paper as you can rather than bunch your data up in one corner. In this case, the interval will be 1. However, we will skip from 0 to 18 by using the zigzag you see just above 0. Notice the labels on both axes. It is important to say what those numbers mean and what those words are. Of course, the whole thing is topped with the title!

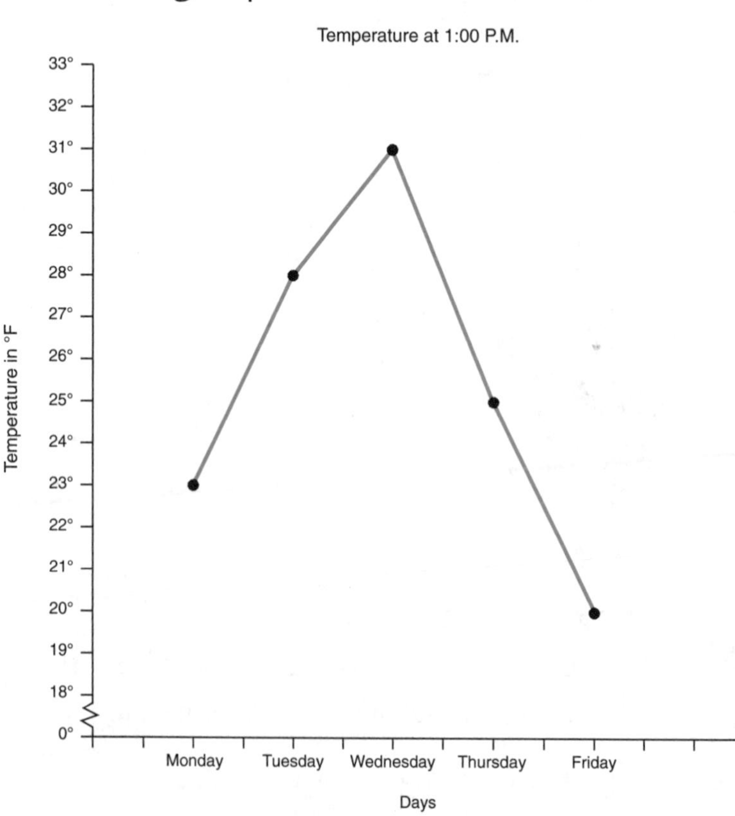

Let's Try It!

Set #3

Mr. Kneeland's class collected newspapers for recycling for six months. The table shows the results of the students' collection in pounds of newspapers.

Month	Newspapers in Pounds
January	56
February	45
March	58
April	50
May	42
June	35

Make a line graph in the grid below. Make sure you write a title, label the axes, and choose a good interval.

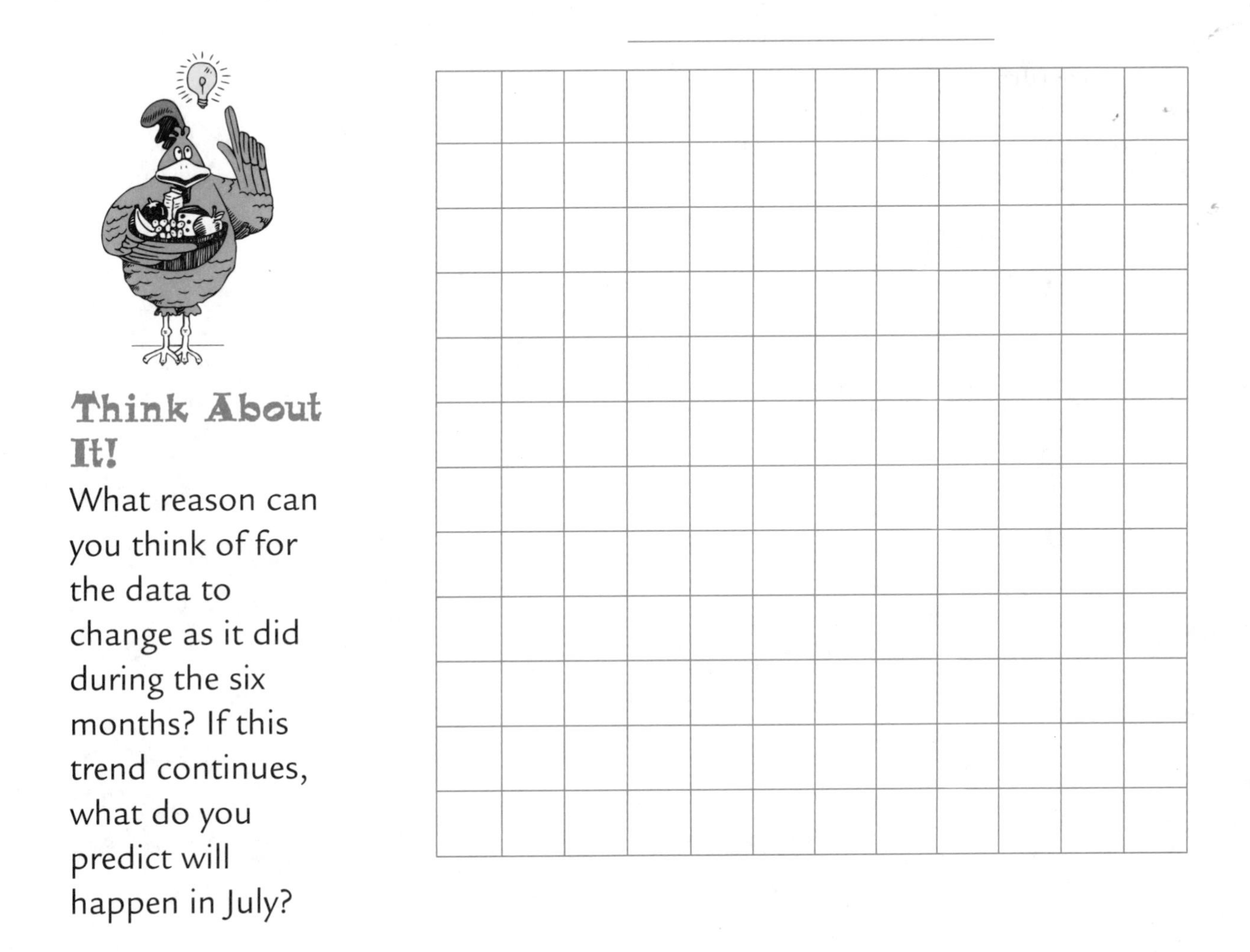

Think About It!

What reason can you think of for the data to change as it did during the six months? If this trend continues, what do you predict will happen in July?

Answers are on pages 184–185.

Chapter 10
HOW MANY PARTS OF THE WHOLE?
A Chapter About Fractions
3/4 1 2/3 2/8
6 5 1/2 1/2

Terms and Definitions

Fraction: A number that names part of a whole or part of a group.

Numerator: The top number of a fraction that tells how many parts are being counted.

Denominator: The bottom number of a fraction that tells how many equal parts are in the whole.

Equivalent fraction: Two or more fractions that name the same amount.

WHAT IS A FRACTION, AND WHERE DOES IT COME FROM?

Fractions are a part of our lives. On the day you were born, you changed the number of people in your family and became a part of that whole unit. You were a fraction of your family!

We hear fraction words in our daily lives. The word *half* is usually used by itself. We might say to a friend, "Would you like half of my sandwich?" What we are really saying is that we are dividing the sandwich into two pieces of the same size.

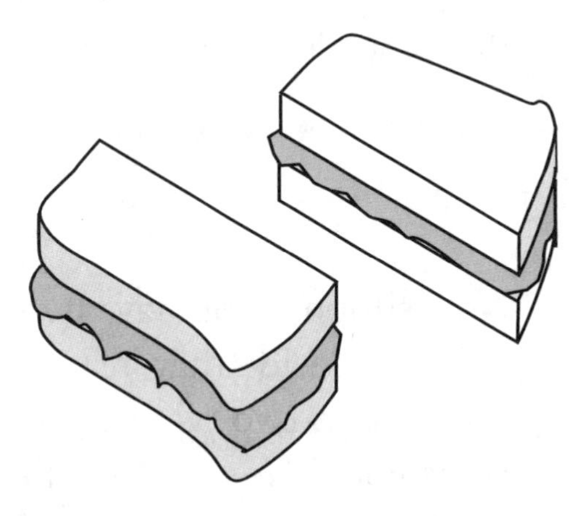

Each piece is then called one half. Most likely, this is the first fraction that was used. People needed fractions when they wanted to divide an object into smaller, equal parts.

IF I SHARE MY COOKIE, DO I GET TO EAT A FRACTION?

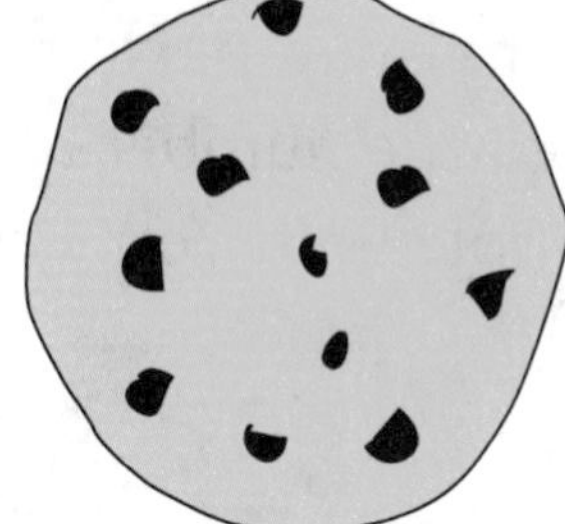

Most if us think of food when we think of fractions. Imagine a giant chocolate chip cookie sitting in front of you right now!

Your best friend likes chocolate chip cookies as much as you do, so you decide to share your cookie. In order to make it fair, you divide the cookie into two equal parts.

If you share one piece with your friend and eat the other piece yourself, you will eat $\frac{1}{2}$ of the cookie. The 1 is called the **numerator**. That's the number on the top of the fraction. It tells how many pieces we have in this fraction. The 2 is the **denominator**, which is the number of pieces the whole has been cut into. That's the number under the line.

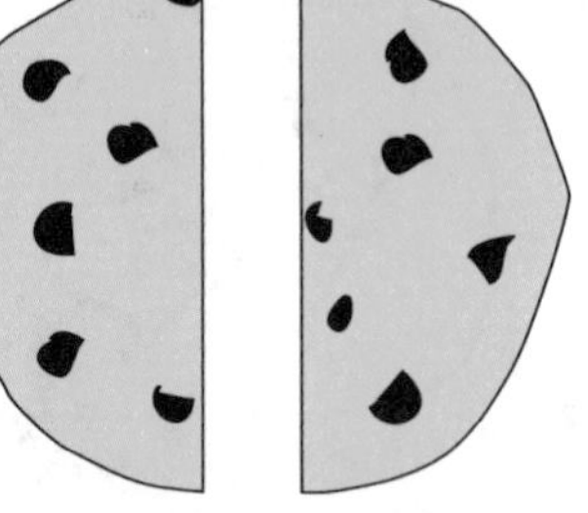

MORE PIECES OF THE COOKIE BUT THE SAME AMOUNT

Now suppose you have two best friends, and you decide that instead of cutting your cookie into two pieces, you will cut it into three equal pieces. One of those pieces is called $\frac{1}{3}$.

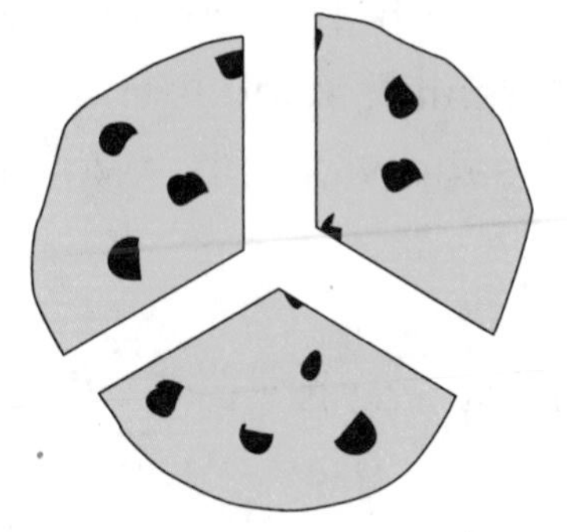

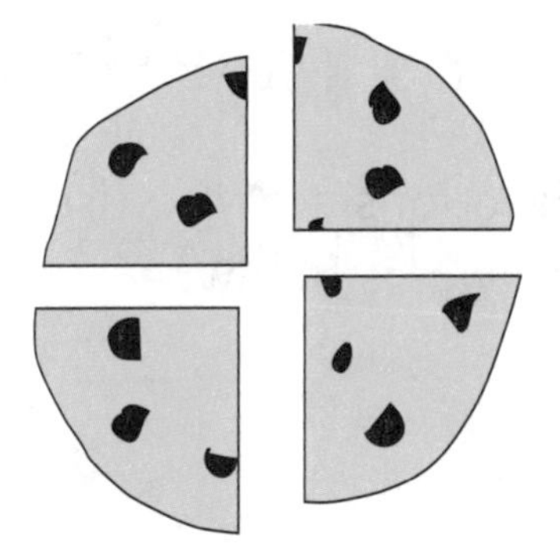

If you are even luckier and have three friends you want to share with, you will divide your cookie into four equal pieces to share. Each of those pieces is called $\frac{1}{4}$.

Do you notice anything about the sizes of the pieces of your cookie? How does the size of $\frac{1}{2}$ compare with $\frac{1}{4}$? Let's take a look at this with a candy bar.

"Look at what's happening to the pieces of candy bar!"

What do you think Pethagerus is noticing about the pieces of the candy bar? If you said that he notices that the pieces are getting smaller, you were right. Each time you divide an object into more equal parts, the pieces get smaller.

Important!

When the numerator and the denominator are the same, the fraction is called **one whole**.

EQUIVALENT FRACTIONS

Fractions are also used to identify parts of a group. Look at the drawings below.

There are 12 shapes in all. Six of them are stars so $\frac{6}{12}$ of the shapes are stars. Four of them are circles so $\frac{4}{12}$ are circles, and two of them are squares so $\frac{2}{12}$ of the group are squares.

Sometimes fractions that are different may have the same value. Here's how it works. Look at this drawing of two of the candy bars.

Can you see that $\frac{1}{2}$ is the same amount as $\frac{2}{4}$? These are called **equivalent fractions**. Look at the fraction bars on the facing page that have been drawn for you. Use them to find other equivalent fractions.

Use a ruler to help you here. Place the ruler vertically on your fraction bars so that the edge is lined up exactly with the $\frac{1}{2}$ line on the second bar. Can you see that your ruler now lines up with $\frac{2}{4}$? What else does it line up with? If you said $\frac{3}{6}$, $\frac{4}{8}$, $\frac{5}{10}$, and $\frac{6}{12}$, you were right. These are all equivalent fractions. The numbers may be different, but they are all the same amount as $\frac{1}{2}$ of your cookie! Look back at the shapes in a group. Do you remember that $\frac{6}{12}$ of the shapes were stars? We can also say that $\frac{1}{2}$ of the shapes are stars.

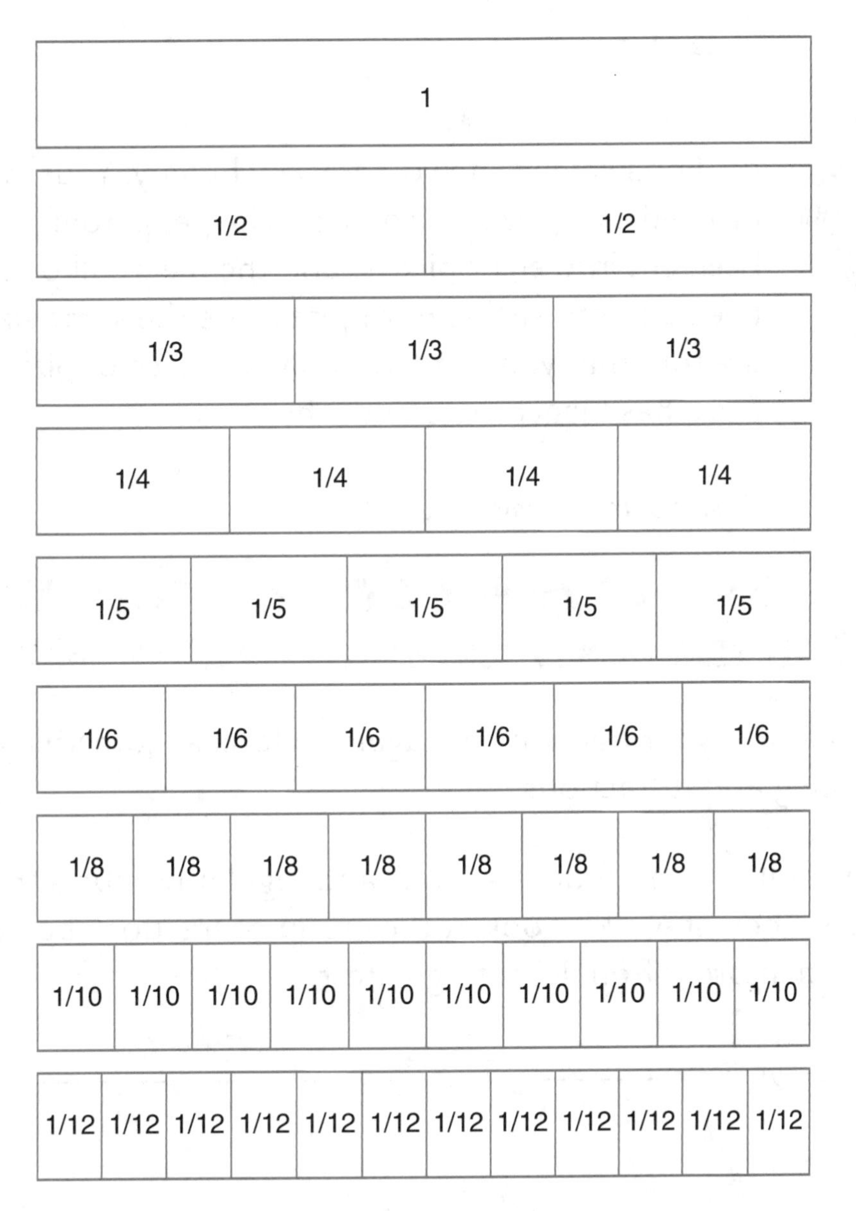

Let's Try It!

Set #1

Choose a fraction that is equivalent to each of the following. Use your fraction bars to help you.

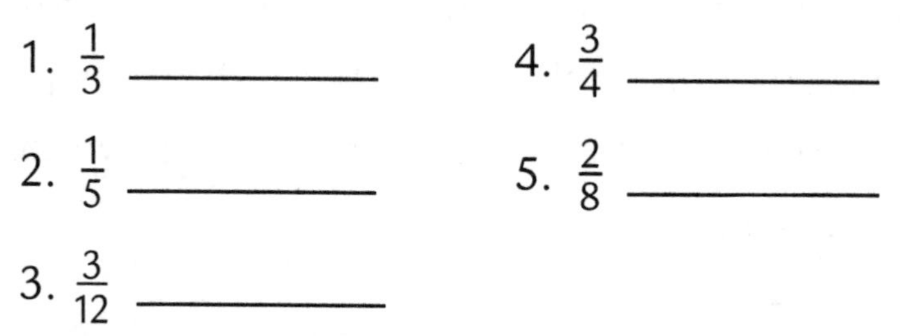

1. $\frac{1}{3}$ ________
2. $\frac{1}{5}$ ________
3. $\frac{3}{12}$ ________
4. $\frac{3}{4}$ ________
5. $\frac{2}{8}$ ________

Think About It!

It's Friday night and you are very hungry. Your family has ordered two pizzas for dinner. The pepperoni pizza has been cut into eight pieces, and the meatball pizza has been cut into sixths. Both pizzas are the same size. You are told that you can have only one slice of pizza. Which pizza has bigger slices and why?

Answers are on page 185.

WHO HAS MORE OF THE COOKIE? ORDERING AND COMPARING FRACTIONS

Let's go back to your fraction bars again to look at something else that is interesting about fractions.

You can compare and order fractions and regular numbers the same way. Here's how it works. Look at the group of fractions below and place them in order from least to greatest.

$\frac{1}{7}, \frac{1}{9}, \frac{1}{2}, \frac{1}{3}$ ______ ______ ______ ______

1

1/2	1/2

1/3	1/3	1/3

1/4	1/4	1/4	1/4

1/5	1/5	1/5	1/5	1/5

1/6	1/6	1/6	1/6	1/6	1/6

1/8	1/8	1/8	1/8	1/8	1/8	1/8	1/8

1/10	1/10	1/10	1/10	1/10	1/10	1/10	1/10	1/10	1/10

1/12	1/12	1/12	1/12	1/12	1/12	1/12	1/12	1/12	1/12	1/12	1/12

"These fractions are neat!"

Since the numerators are all the same, we only need to look at the denominators. If you use your fraction bars and what you know about the size of the denominators, this is an easy job.

OK! The smallest pieces have the highest denominators, so that means that the order is:

$\frac{1}{9}, \frac{1}{7}, \frac{1}{3}, \frac{1}{2}$

How did you do?

Let's try some fractions with the same denominator. Can you put the following fractions in order from least to greatest?

$\frac{6}{8}, \frac{3}{8}, \frac{5}{8}, \frac{7}{8}, \frac{1}{8}$

This time the denominators are alike, so just use the numerators to place these in order. Start with the least numerator and away you go!

$\frac{1}{8}, \frac{3}{8}, \frac{5}{8}, \frac{6}{8}, \frac{7}{8}$

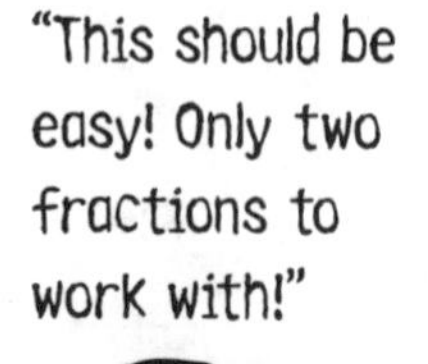

You can also compare just two fractions at a time using < (less than), > (greater than), or = (equal to).

Here's how this works. We'll start with the two fractions you see below.

$\frac{1}{2} \qquad \frac{1}{5}$

We know by looking at them that they are not equal. They have the same numerator but different denominators. Since halves are larger pieces than fifths, we can write our answer like this:

$\frac{1}{2} > \frac{1}{5}$

When the numerators are different but the denominators are the same, we might have a problem like this:

$\frac{2}{5} \quad \frac{3}{5}$

Since 2 is less than 3, our answer should look like this:

$\frac{2}{5} < \frac{3}{5}$

Let's Try It!

Set #2

Order these fractions from least to greatest.

1. $\frac{7}{9}, \frac{3}{9}, \frac{2}{9}, \frac{5}{9}$ ____ ____ ____ ____

2. $\frac{3}{12}, \frac{5}{12}, \frac{2}{12}, \frac{8}{12}, \frac{1}{12}$ ____ ____ ____ ____ ____

3. $\frac{3}{8}, \frac{3}{4}, \frac{3}{5}, \frac{3}{9}, \frac{3}{3}$ ____ ____ ____ ____ ____

Compare these fractions using $<$, $>$, or $=$.

4. $\frac{6}{8}$ ____ $\frac{3}{4}$

5. $\frac{1}{3}$ ____ $\frac{1}{4}$

6. $\frac{2}{5}$ ____ $\frac{2}{3}$

Think About It!

How would you go about comparing $\frac{2}{3}$ with $\frac{3}{5}$?

Answers are on page 186.

LET'S PUT THE PIECES TOGETHER—ADDING AND SUBTRACTING

When you were younger, the first operation you learned was addition. Now you can learn to add and subtract fractions, too! You use what you already know about fractions to help you. Think back to the cookies we divided early in this chapter. If the cookie is divided into four equal parts, we know that each of those parts is $\frac{1}{4}$. If you eat two of those pieces and your friend eats one of them, you can figure out what fraction of the cookie was eaten by adding $\frac{2}{4}$ and $\frac{1}{2}$. Here's how it looks:

$$\frac{2}{4} + \frac{1}{4} = \frac{3}{4}$$

Why do we add the numerator and not the denominator? If you said the denominator does not change because the size of the pieces does not change, you were right. Think of the denominator as a name. If you add two jellybeans and three jellybeans, you will have a sum of five jellybeans—not five marshmallows.

Subtraction works the same way. If you have $\frac{3}{5}$ of a cookie and eat $\frac{2}{5}$ of it, you will have $\frac{1}{5}$ left. It looks like this:

$$\frac{3}{5} - \frac{2}{5} = \frac{1}{5}$$

Let's Try It! Set #3

Add or subtract.

1. $\frac{7}{8} - \frac{4}{8} =$ ________
2. $\frac{3}{10} + \frac{4}{10} =$ ________
3. $\frac{4}{9} + \frac{2}{9} =$ ________
4. $\frac{11}{12} - \frac{7}{12} =$ ________
5. $\frac{1}{6} + \frac{4}{6} =$ ________

Think About It!

What is special about the sum of this problem?

$\frac{5}{8} + \frac{3}{8}$

Answers are on page 186.

Chapter 11
WHAT IS LIKELY
TO HAPPEN?
A Chapter About Probability

Terms and Definitions

Probability: The chances of an event occurring.

Certain: An event that will always happen.

Impossible: An event that will never happen.

Likely: An event that may happen but is not certain.

Unlikely: An event that may not happen but is not impossible.

Equally likely: Events having the same chance of happening.

IS IT CERTAIN OR IMPOSSIBLE?

It is lunchtime, and you turn to your friend and say, "I probably have a sandwich today for lunch." Your friend says, "You always have a sandwich!" In this chapter, we are going to learn about **probability**, or the chances of an event occurring. Actually, probability is found in a very small space in terms of numbers. An event that will always happen is **certain**. It has a probability of 1. An event is **impossible** if it will never happen. It has a probability of 0. All the rest of probability falls between 0 and 1. In other words, most of probability is in the form of a fraction. We'll get to that later in the chapter.

Can you think of some events that are certain and have a probability of 1? It is certain that you get wet when you go swimming. It is certain that snow feels cold on your nose.

Can you think of some events that are impossible and have a probability of 0? It is impossible to see a real dinosaur walk down the street. It is impossible for a marble to become a cookie.

LIKELY AND UNLIKELY EVENTS

Now let's talk about the part of probability that is between certain and impossible. Sometimes an event is **likely** because it is not certain but very close to it! Other times an event is **unlikely** because it isn't impossible but very close to that!

Can you think of some likely events? Perhaps it is likely that you will get homework tonight. It is likely that you will see your friends at school.

Can you think of some unlikely events? It is unlikely that you will stay home from school due to snow if you live in Florida. It is unlikely that the temperature will be 75°Farenheit in January if you live in Maine.

Let's Try It!

Set #1

Identify each of these events by writing certain, impossible, likely, or unlikely.

1. The day after Monday will be Saturday. ____________
2. The sun will rise tomorrow morning. ____________
3. You will eat lunch today. ____________
4. All the teachers will be absent from school tomorrow. ____________
5. You will wear shoes to school today. ____________

Think About It!

List some events. Think of one that is certain, one that is impossible, one that is likely, and one that is unlikely.

Answers are on page 186.

EQUALLY LIKELY EVENTS

Have you ever played a game with a spinner? We are going to use some of the spinners below.

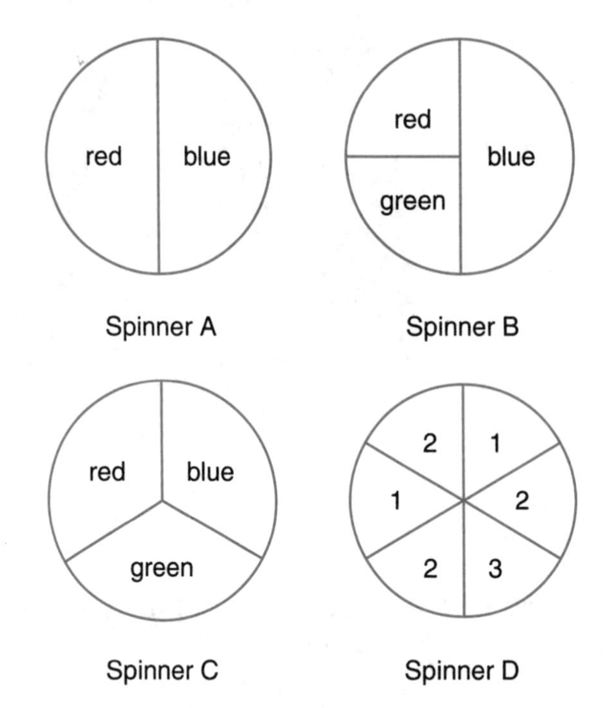

Spinner A Spinner B

Spinner C Spinner D

Look at Spinner A for a moment. Do you notice that both sections are the same size? When that happens, we can say that it is **equally likely** that red or blue will be the result of a spin. Can you find another spinner with a result that is equally likely? If you said Spinner C, you were right. Each of the three colors is taking up the same amount of space on the spinner.

If you need to spin red to win the game you are playing, which spinner should you choose? If you said Spinner A, you were right. On Spinner A red has the same chance as blue. On Spinner B, red does not have a good chance. On Spinner C, red has the same chance as blue and green.

USING FRACTIONS

Remember when fractions were mentioned in the beginning of the chapter as a part of probability? Here's how it works.

Look at Spinner A. There are two sections of the same size. One is red and the other is blue. The chance of spinning blue is one chance in two or $\frac{1}{2}$. The chance of spinning red is also $\frac{1}{2}$.

Now look at Spinner C. The three sections are the same size, so the chance of spinning red is $\frac{1}{3}$. It is also $\frac{1}{3}$ for spinning blue and $\frac{1}{3}$ for spinning green.

Our friend asks a good question. The six sections are the same, but the numbers are not all the same. What do you think the probability of spinning a 3 is? If you said one chance in six or $\frac{1}{6}$, you were right. What is the probability of spinning a 2? If you said three chances in six or $\frac{3}{6}$, you were right. What is the probability of spinning a 1? If you said two chances in six or $\frac{2}{6}$, you were right.

Did you notice that Spinner B was different from the others? The sections are not all the same size. This makes for an unfair game. If you wanted to win, which color do you think would most likely result? If you said blue, you were right. The blue space is twice the size of either red or green.

MAKING PREDICTIONS

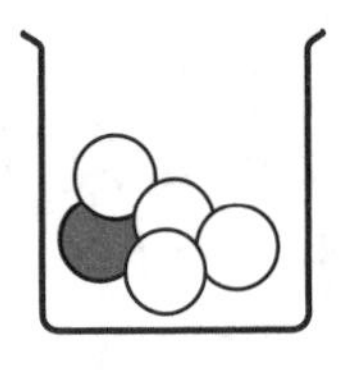

Another part of probability is predicting outcomes or events. Look at the container of marbles. Can you predict what will happen if you take a marble out of the container without looking? Here's where you use what you know about probability. Since there are four white marbles and only one black marble, it is more likely that you will pull out a white marble. Your chances are 4 out of 5. If you need to pull a white marble to win the game, probability is on your side. As a matter of fact, if you pull a marble out of the container, record the results, and place it back in the container before pulling another marble out of the container, your results should look just like these chances. In other words, for every five marbles you pull out of the container you should see only one black.

Let's Try It!

Set #2

Write the probability as a fraction for each of the following:

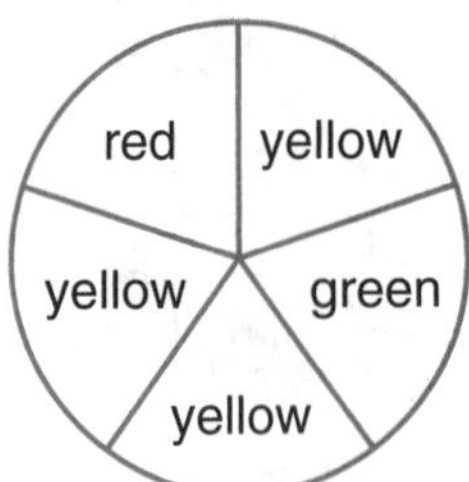

1. What is the probability of spinning a yellow on this spinner?

   ____

2. What is the probability of rolling a five on this regular die?

   ____

3. What is the probability of rolling an even number on that same die?

   ____

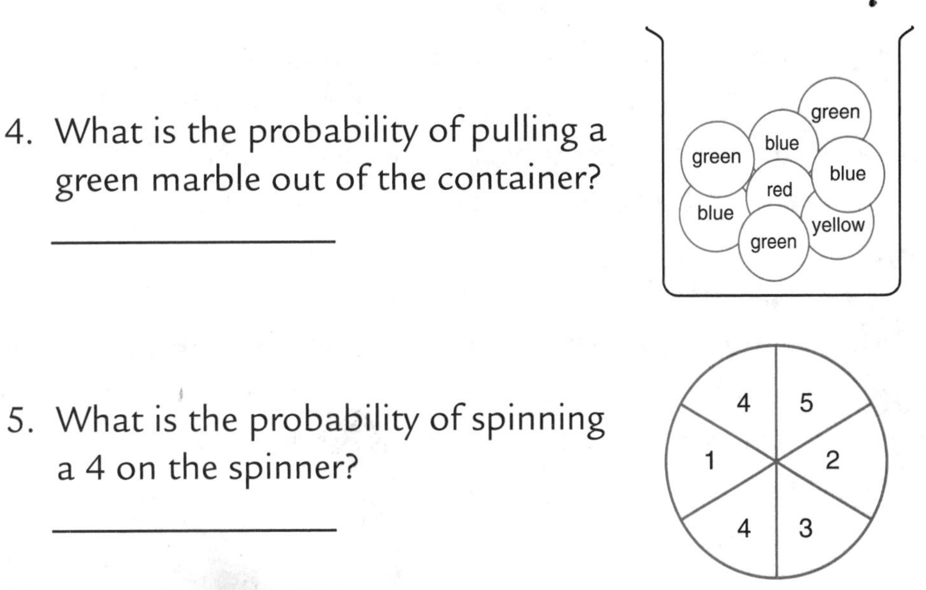

4. What is the probability of pulling a green marble out of the container?

    ____

5. What is the probability of spinning a 4 on the spinner?

    ____

Think About It!

What is the probability of spinning a blue on the spinner? Explain your answer.

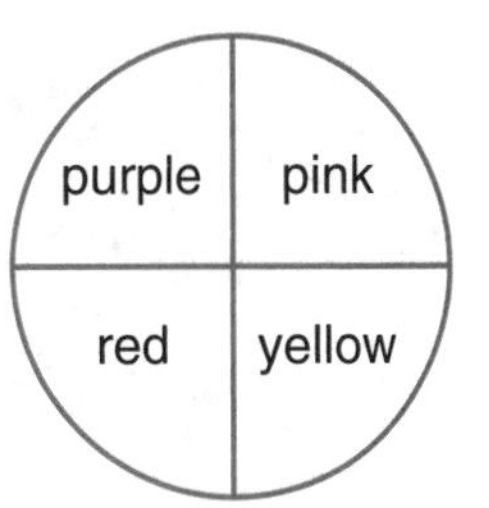

Answers are on pages 186–187.

ORGANIZING RESULTS

OK! So you want to prove to yourself this probability thing! How do you organize results? Let's go back to the bag of marbles. It has eight marbles in it. Three of them are green, four of them are blue, and one is white. Set up a table like the one below.

Color	Tally	Total
Green		
Blue		
White		

Begin by pulling out a marble without looking. Record its color using a tally mark on the chart. Put the marble back in the bag. Repeat this process as many times as you wish. The more times you pull out a marble and record it, the better your results will be.

Careful!

When you record tally marks, make four marks and then bundle them together with the fifth one across them. This way you can count them as easily as you count by fives.

Here is a sample table.

Color	Tally	Total
Green	卌 |||	8
Blue	卌 卌	10
White	||	2

Pethagerus would like you to try this out! Use two dice and roll them both. Add the two numbers you have rolled and record the results on the table below.

Sum	2	3	4	5	6	7	8	9	10	11	12
Tally											
Total											

What did you predict? Did you expect any one sum to appear more than others? Let's take a look at this by taking each sum.

The black numbers are one die and the red numbers are the other die.

2 1 + 1
3 1 + 2, 2 + 1
4 1 + 3, 2 + 2, 3 + 1
5 1 + 4, 2 + 3, 3 + 2, 4 + 1
6 1 + 5, 2 + 4, 3 + 3, 4 + 2, 5 + 1
7 1 + 6, 2 + 5, 3 + 4, 4 + 3, 5 + 2, 6 + 1
8 2 + 6, 3 + 5, 4 + 4, 5 + 3, 6 + 2
9 3 + 6, 4 + 5, 5 + 4, 6 + 3
10 4 + 6, 5 + 5, 6 + 4
11 5 + 6, 6 + 5
12 6 + 6

The more rolls you record, the more you will see the sums with more combinations turn up.

What happened when you tried it?

Chapter 12
GOT PROBLEMS?
A Chapter About Problem Solving

Terms and Definitions

Strategy: A method chosen to solve a problem.

Strategies in This Chapter:

Draw a picture

Make a list

Find a pattern

Act it out

Make a table

Guess and check

Work backward

Solve a simpler problem

WHAT ARE THESE THINGS CALLED STRATEGIES?

Uh oh! It's Friday again! That most likely means it is Problem Solving Day! For many of you, this might be enough to strike fear into your hearts. Not to worry! Pethagerus is here to help you!

"This problem solving is fun! Let's do it!"

You have probably heard your teacher talk about **strategy**. Perhaps you have wondered what that is all about! There are a number of problem-solving strategies that we are going to explore together. Once you feel comfortable with them, you will find that problem solving is much easier.

DRAW A PICTURE

The first strategy is to **draw a picture**. This strategy seems simple enough. Look at the problem below and think about the picture you might draw.

Sasha has just finished drawing a picture of a figure skater. She wants to put a piece of ribbon around the edge of her picture. The picture is a rectangle that has two long sides measuring 8 inches each and two short sides measuring 5 inches each. How long is the piece of ribbon that Sasha should use to go around the picture once?

5 in.

8 in. 8 in.

5 in.

Look at the picture drawn for you. The rectangle has been labeled so that you can see how long each side is.

In order to find out how much ribbon is needed to go around the picture, you can add 8 inches + 8 inches + 5 inches + 5 inches to get a sum of 26 inches. Sasha needs 26 inches of ribbon to go around the edge of her picture once.

Let's Try It!

Set #1

Put this strategy to work on these problems.

1. Alvaro is planning his display for the science fair. The display board is 30 inches tall and 24 inches wide. He has four pictures, and each one is 12 inches tall by 10 inches wide. Will they fit on Alvaro's display board?

   ____

2. Jenna's dad is putting up a fence around their pool. The pool is 20 feet long and 15 feet wide, and the fence will be 5 feet away from the pool on all sides. How many feet wide and how many feet long will the fence be around Jenna's pool?

   ____

3. Greg wants to send two of his framed pictures to his grandmother. One picture is 18 inches long and 12 inches wide, and the other picture is a 15 inch square. How big does the box have to be for both pictures to fit?

   ____

Answers are on pages 187–188.

MAKE A LIST

The next strategy we will work with is called **make a list**. Sometimes a problem requires you to figure out more than one answer. It's a good idea to make an organized list so you are sure to get all of the answers. Here's how it might look.

How many different three-digit numbers can you make using 1, 2, and 3 every time?

If we make a list of the numbers and organize our answer, we will be sure to have it all. Begin with all the three-digit numbers that start with 1.

123
132

Then make all the three-digit numbers you can that begin with 2.

213
231

Now make all the three-digit numbers you can that begin with 3.

312
321

We put the lists together and here is our answer.

123	213	312
132	231	321

Let's Try It!

Set #2

1. Rebecca is going to camp. She has packed a pink shirt and a blue shirt. She has packed brown shorts and white shorts. She also has packed a yellow cap and a green cap. If she wears one of each to make an outfit, how many different outfits has she packed?

   ____

2. Kathryn is making lunch. She looks in the refrigerator and finds ham and salami. She also finds some mustard and mayonnaise. There is a roll or rye bread for Kathryn to choose from. If she makes a sandwich using one kind of bread, one filling, and one spread, how many different sandwiches can Kathryn make?

   ____

3. Donna loves ice cream cones. She has chocolate or vanilla ice cream. She can have sprinkles or a cherry on it and either a sugar cone or a waffle cone. If she has one kind of ice cream, one topping, and one kind of cone, how many different ice cream cones can Donna make?

   ____

Answers are on page189.

FIND A PATTERN

When you were in kindergarten you were most likely introduced to making patterns with all kinds of blocks. Now that you are older, you have the opportunity to look for patterns in numbers. The next strategy is called **find a pattern**. Here's how it works.

Susan was reading a sign at the movie theater. It said that on Monday the fifth person in line would get a free drink at the snack bar. On Tuesday the tenth person on line would get the free drink and on Wednesday the 15th person on line would get the free drink. If this pattern continues, which person will get the free drink on Saturday?

Let's make a T-chart to help us find this pattern.

Day	Person in Line
Monday	5th
Tuesday	10th
Wednesday	15th
Thursday	20th
Friday	25th
Saturday	30th

So the answer is the 30th person in line on Saturday will get the free drink at the snack bar.

Let's Try It!

Set #3

1. Nora is doing some chores in her home. She receives $1.50 at the end of the first week, $2.00 at the end of the second week, $2.50 at the end of the third week, and $3.00 at the end of the fourth week. If this rate continues, what will Nora earn at the end of the sixth week?

2. During July, there are concerts in the park. So far, the concerts have been on July 2, July 9, and July 16. If this pattern continues, on what date will the fifth concert be given?

3. Brenda is making a winter arrangement of pinecones. The first row has five pinecones, the second has eight pinecones, and the third has 11 pinecones. If this pattern continues, how many pinecones will Brenda use in the fifth row?

Answers are on pages 189–190.

ACT IT OUT

The next strategy, called **act it out**, is a lot of fun because you can actually get friends to help you solve a problem by moving around. Here's how this works.

Kelly, Gabe, Hope, and Maria are standing in line for the movies. Kelly is between Maria and Hope. Hope is not last. Maria is first. In what order are the four friends waiting in line for the movies?

In order to solve this problem, you can get three friends to help you act this out together. Make a nametag for each person in the problem for you and your friends to hold. Now stand in line and change positions until you have Kelly between Maria and Hope. Then move to make Maria first. If Hope is not last, then Gabe must be last, so the order is as follows:

Maria,
Kelly,
Hope,
Gabe

You can also solve this problem by acting it out using just the nametags and moving them into the proper position.

Let's Try It!

Set #4

1. Millie is having a pizza party. Her mother has tables that seat five people and tables that seat six people. She wants all the tables filled. There will be 23 people, including Millie, at the party. How many tables seating five and how many tables seating six will Millie's mother need?

Hint: You can act this out by cutting out strips of paper, 5 or 6 of them. Group them together until you find a combination of tables adding up to 23.

2. Livvie is folding six t-shirts to put on a shelf. She puts them in two stacks of three t-shirts. The yellow shirt is to the right of the pink shirt. She puts the white shirt under the pink shirt. The red shirt is to the right of the white shirt, and the yellow shirt is between the blue shirt and the red shirt. She also has a green t-shirt. Where does Livvie put each of her t-shirts?

Hint: Write thee color words in this problem on separate scraps of paper. Move them around to fit the problem.

3. Allan is packing lunches in backpacks for a hiking trip. The backpacks can hold five lunches or four lunches, and each backpack must be full. Allan has 17 lunches to pack. How many backpacks of each kind can Allan pack so that each backpack is full and there are no lunches left over?

Hint: Act this out by writing 4 or 5 scraps of paper. Group them together until you get a sum of 17.

Answers are on pages 190–191.

MAKE A TABLE

We turn now to a strategy called **make a table**.

No, not that kind of table, Pethagerus! Here's how it works.

Zoë has a kitten that weighs 2 pounds and a puppy that weighs 7 pounds. Each pet is gaining one pound each month, and soon the puppy will weigh twice as much as the kitten. How much will both animals weigh then?

Set up your table to look like this.

Kitten	2 pounds	3 pounds	4 pounds	5 pounds
Puppy	7 pounds	8 pounds	9 pounds	10 pounds

When the kitten weighs 5 pounds, the puppy will weigh twice that amount or 10 pounds.

This can organize your work and help you think through the problem.

Let's Try It!

Set #5

1. Anne is planning her summer hiking trips. She has recorded the length of several local trails and would like to hike them all, beginning with the shortest. The Hound's Tooth Trail is eight miles long, the Quiet Elm Trail is 10 miles long, the Elegant Fir Trail is six miles long and the Long Beach Trail is nine miles long. In what order should Anne hike these trails?
2. Jack and Duncan love to play baseball. They visit the batting cage to practice their hitting. Jack goes every three days and Duncan goes every four days. On what day will Jack and Duncan both be at the batting cage?
3. A turtle and a frog are headed for a pond that is 32 inches away from them. The turtle is able to double his distance every hour. The frog jumps six inches each hour and then takes a rest. Which animal will get to the pond first?

Answers are on pages 191–192.

GUESS AND CHECK

"How can I get the answer if I just guess it?"

We move on to a strategy called **guess and check**. We solve problems by guessing first to see how close we can come. If the answer is too much, we take some away. If the answer is too little, we add some. When the answer is right, we will know it. Here's how it works.

Orrin and Noel are comparing their collections of baseball cards. Together they have 51 cards. Orrin has 15 more cards than Noel has. How many baseball cards does each boy have? Let's see how this one goes. If we divide 51 by 2 we get 25 with a remainder of 1. That won't work at all! We could subtract 15 from 51 to get a difference of 36. Does that work out? If Orrin has 36 cards then Noel has 15.

$$36 - 15 = 21$$

We are getting closer. Now let's guess that Orrin has 35 cards. Noel would have 16.

$$35 - 16 = 19$$

We are getting closer! Take another away from Orrin so that he now has 34. Noel will have 17.

$$34 - 17 = 17$$

We are *very* close! Take one more away for Orrin to give him 33 and Noel 18. Let's check it!

$$33 - 18 = 15$$

There it is!

Let's Try It! Set #6

1. Clara had a party for her friends and made 30 cookies. There were more than five people at the party and the cookies were handed out equally. After the party Clara noticed that there were six cookies left. How many people were at Clara's party?

2. There were 24 black and red marbles in a bag. There were twice as many red marbles as black marbles. How many of each were in the bag?
3. Bethany had 47 books to put on shelves. She had five books left over and one more shelf than the number of books on those shelves. How many shelves did Bethany use? How many books were on each shelf?

Answers are on pages 192–193.

WORK BACKWARD

Have you ever tried walking backward? It's fun until you bump into someone that you couldn't see. There is a problem-solving strategy called **work backward**.

Here's how it works.

Nancy spent $14. She then had one ten-dollar bill, one five-dollar bill, and two one-dollar bills in her pocket. How much money did Nancy have before she spent some?

To solve this problem by working backward, add the money Nancy had in her pocket.

$$\$10 + \$5 + \$2 = \$17$$

Now let's add to that the money she spent.

$$\$14 + \$17 = \$31$$

Nancy had $31 before she spent any of it.

Let's Try It!

Set #7

1. Ron practices the piano on weekends for 45 minutes. If he finishes practicing at 12:30 P.M., what time does he begin?
2. Ona buys a guitar case and uses a one hundred-dollar bill. With the change, she buys herself some lunch for $10 and has $17 left. How much money did the guitar case cost?
3. Arny made a sandwich with two 1-ounce slices of bread, 4 ounces of cheese, and some salami. Before he ate the sandwich, Arny weighed it. If the sandwich weighed 11 ounces, how much did the salami weigh?

Answers are on page 193.

SOLVE A SIMPLER PROBLEM

The last strategy we will explore is called **solve a simpler problem**. What does that mean? If you have a problem that has either lots of information or bigger numbers, it can help a lot if you change the problem or the numbers to make it simpler. That way you can make a plan. When you have that plan in mind, then you can use the original numbers or the original information.

"What a great idea!"

Let's see how it works.

The Smithdale Country Club is buying a new fence. One yard costs $29.95. Ms. Olga, the club president, says that 45 yards are needed. How much will the new fence cost?

Solving this problem involves using large numbers. The money is not on the dollar and the 95¢ makes it more complicated. Let's make this simpler by using \$30 and 50 yards. To solve this simpler problem we multiply the \$30 times the 50 yards.

$$\$30 \times 50 \text{ yards} = \$1{,}500$$

Check to see that this answer makes sense for the problem. Then find the actual answer by doing the following:

$$\$29.95 \times 45 \text{ yards} = \$1{,}347.75$$

This answer is close to the answer for the simpler problem and a little less than that answer. That makes sense because we made both numbers larger to make it simpler.

Let's Try It!

Set #8

1. Jamie, Emma, Marley, and Selena are preparing a dance routine for the talent show. How many different ways can they stand in line for their dance?
2. The Westerhoff family is going to see a show. Tickets for adults are \$8.78 each and the children's tickets are \$5.49 each. There are two adults and three children. How much will the Westerhoffs pay for the show?
3. Each member of Mrs. Deger's art club made 55 drawings for the local hospital. If there are 47 members in Mrs. Deger's art club, how many drawings did they make in all?

Answers are on pages 193–194.

Answer Key

CHAPTER 1

Answers to Let's Try It!

Set #1 page 4

1. 8:00
2. 3:00
3. 10:00
4. 1:00
5. 6:00

Answer to Think About It!

12:00

Answers to Let's Try It!

Set #2 page 5

1. 7:30
2. 2:30
3. 10:30
4. 5:30
5. 9:30

Answer to Think About It!

The minute hand is on the 6 and the hour hand is halfway between the 6 and the 7.

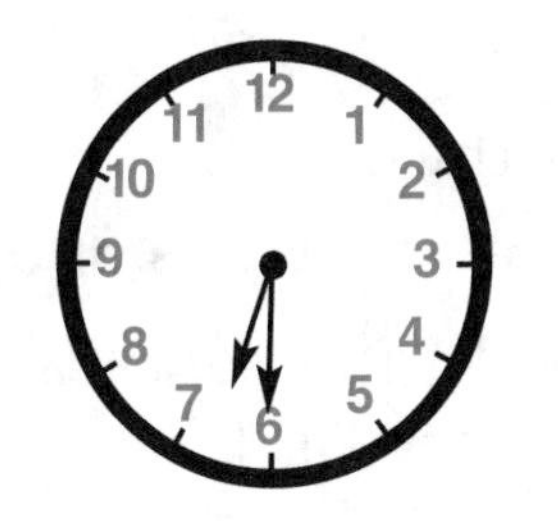

Answers to Let's Try It!

Set #3 page 6

1. 9:35
2. 1:25
3. 11:10
4. 7:50
5. 3:40

Answer to Think About It!

There are 5 minutes left in the hour.

Answers to Let's Try It!

Set #4 page 9

1. A.M.
2. P.M.
3. P.M.
4. P.M.
5. A.M.

Answer to Think About It!

A.M. changes to P.M. at 12:00, and that time is called noon.

Answer to Think About It!
page 9

No, you need other clues to help you like these:

Is the sun out?

Is it night with a sky full of stars and maybe the moon?

Am I eating breakfast or dinner?

Some digital clocks will show a little A.M. or P.M.

Answers to Let's Try It!
Set #5 page 12

1. Sunday
2. There are five Saturdays in May.
3. There are four Tuesdays in May.
4. There are 2 weeks and 2 days.
5. Tuesday

Answer to Think About It!

No, because the longest month has 31 days. These are placed in groups of seven for each week. Four groups of seven days are 28, so there are not enough days to have more than five in a month.

Answers to Let's Try It!
Set #6 page 14

1. 40 minutes
2. 3 hours and 25 minutes
3. 2 hours and 45 minutes
4. 2 hours and 5 minutes
5. 55 minutes

Answer to Think About It!

A.M. and P.M. are important because they give us a general time of day. If you want to watch an important television program at 8:00 you also need to know if it is 8:00 A.M. or 8:00 P.M. You would miss it if you tuned in at 8:00 P.M. and discovered it was already shown at 8:00 A.M.

CHAPTER 2

Answers to Let's Try It!
Set #1 page 18

1. The value of the 3 is 300, the value of the 9 is 90, and the value of the 2 is 2.
2. The value of the 4 is 400, the value of the 7 is 70, and the value of the 9 is 9.
3. The value of the 5 is 500, the value of the 8 is 80, and the value of the 1 is 1.

4. The value of the 6 is 600, the value of the 3 is 30, and the value of the 5 is 5.
5. The value of the 2 is 200, the value of the 4 is 40, and the value of the 7 is 7.

Answer to Think About It!

When there is a 0 in any place there is no value for the place, but it is very important. In the number 405, if the zero was left out the number would be only 45. That makes a big difference!

Answers to Let's Try It!

Set #2 page 21

1. 495 < 594
2. 301 > 298
3. 732 < 742
4. 900 > 899
5. 678 = 678

Answer to Think About It!

The ones place is only important if all the other places are the same. Here is an example: 467 < 468.

Answers to Let's Try It!

Set #3 page 24

1. five thousand nine hundred fifty-two
2. sixty-seven thousand, four hundred thirty-two
3. four hundred seventy-nine thousand, twenty-five
4. nine thousand, nine hundred ninety-nine
5. eight hundred twenty-four thousand, five hundred sixty-three

Answer to Think About It!

The next number is 1,000,000.

Answers to Let's Try It!

Set #4 page 27

1. 56 rounded to the nearest ten is 60.
2. 25 rounded to the nearest ten is 30.
3. 378 rounded to the nearest ten is 380.
4. 1,541 rounded to the nearest ten is 1,540.
5. 89,659 rounded to the nearest ten is 89,660.

Answer to Think About It!

Yes, it is a ten because if you counted by tens you would eventually get to 1,000, *and* 1,000 has a zero in the ones place. All tens have a zero in the ones place.

Answers to Let's Try It!

Set #5 page 29

1. 129 rounded to the nearest hundred is 100.
2. 451 rounded to the nearest hundred is 500.
3. 764 rounded to the nearest hundred is 800.
4. 3,897 rounded to the nearest hundred is 3,900.
5. 15,678 rounded to the nearest hundred is 15,700.

Answer to Think About It!

49 rounded to the nearest hundred is 0.

CHAPTER 3

Answers to Let's Try It!

Set #1 page 35

1. 50 + 60 = 110
   2 + 3 = 5
   110 + 5 = 115
2. 30 + 50 = 80
   8 + 1 = 9
   80 + 9 = 89
3. 50 + 70 = 120
   5 + 3 = 8
   120 + 8 = 128
4. 200 + 500 = 700
   80 + 10 = 90
   1 + 7 = 8
   700 + 90 + 8 = 798
5. 300 + 600 = 900
   20 + 60 = 80
   7 + 2 = 9
   900 + 80 + 9 = 989

Answer to Think About It!

When your mental math problem contains one number that is greater by a place value, just add the place value along with your answer. Here's an example.
326 + 52

First add the tens.
20 + 50 = 70

Then add the ones.
6 + 2 = 8

Now add 300 + 70 + 8 = 378

Answers to Let's Try It!

Set #2 page 37

1. 59 + 0 = 59   I
2. 34 + 62 = 62 + 34   C
3. 0 + 398 = 398   I
4. (3 + 5) + 2 = 3 + (5 + 2)   A
5. 857 + 245 = 245 + 857   C

Answer to Think About It!

Yes, it is possible to use more than one property in a single equation. Here is an example:

71 + 0 = 0 + 71

The identity property and the commutative property are both used.

Here's another example:

$$\begin{aligned}(4 + 1) + 5 &= (5 + 1) + 4\\ 5 + 5 &= 6 + 4\\ 10 &= 10\end{aligned}$$

This equation used both the associative and commutative properties.

Answers to Let's Try It!

Set #3 page 40

1.
$$\begin{array}{r} 1\,1 \\ 5{,}793 \\ +\ 7{,}351 \\ \hline 13{,}144 \end{array}$$

2.
$$\begin{array}{r} 11 \\ 459 \\ 731 \\ +\ 865 \\ \hline 2{,}055 \end{array}$$

3.
$$\begin{array}{r} 1\ \ 1 \\ 9{,}834 \\ +\ 8{,}609 \\ \hline 18{,}443 \end{array}$$

4.
$$\begin{array}{r} 21 \\ 693 \\ 452 \\ +\ 975 \\ \hline 2{,}120 \end{array}$$

5.
$$\begin{array}{r} 2\ 11 \\ 6{,}832 \\ 7{,}895 \\ +\ 5{,}649 \\ \hline 20{,}376 \end{array}$$

Answer to Think About It!

When you add numbers together you always get a greater sum. Look at the Let's Try It! questions you just answered. Do you notice that every sum is larger than any of the addends you added?

CHAPTER 4

Answers to Let's Try It!

Set #1 page 45

1.
$$\begin{array}{rrr} 7 & 12 & 11 \\ \cancel{8} & \cancel{3} & \cancel{1} \\ -5 & 7 & 6 \\ \hline 2 & 5 & 5 \end{array}$$

2.
$$\begin{array}{rrr} 8 & 15 & \\ \cancel{9} & \cancel{5} & 6 \\ -3 & 9 & 5 \\ \hline 5 & 6 & 1 \end{array}$$

3.
$$\begin{array}{rrr} 5 & 16 & 14 \\ \cancel{6} & \cancel{7} & \cancel{4} \\ -4 & 8 & 5 \\ \hline 1 & 8 & 9 \end{array}$$

4.
$$\begin{array}{rrr} 6 & 13 & 13 \\ \cancel{7} & \cancel{4} & \cancel{3} \\ -2 & 6 & 6 \\ \hline 4 & 7 & 7 \end{array}$$

5.
$$\begin{array}{rrr} 4 & 17 & \\ \cancel{5} & \cancel{7} & 2 \\ -1 & 8 & 1 \\ \hline 3 & 9 & 1 \end{array}$$

Answer to Think About It!

When numbers are subtracted, the difference is always a smaller number.

Answers to Let's Try It!

Set #2 page 47

1.
$$\begin{array}{rrr} & 9 & \\ 8 & \cancel{10} & 16 \\ \cancel{9} & \cancel{0} & \cancel{6} \\ -5 & 9 & 8 \\ \hline 3 & 0 & 8 \end{array}$$

2.
$$\begin{array}{rrr} & 9 & \\ 7 & \cancel{10} & 11 \\ \cancel{8} & \cancel{0} & \cancel{1} \\ -2 & 7 & 3 \\ \hline 5 & 2 & 8 \end{array}$$

3.
$$\begin{array}{rrr} & 9 & \\ 3 & \cancel{10} & 13 \\ \cancel{4} & \cancel{0} & \cancel{3} \\ -2 & 8 & 4 \\ \hline 1 & 1 & 9 \end{array}$$

4.
$$\begin{array}{rrr} & 9 & \\ 5 & \cancel{10} & 14 \\ \cancel{6} & \cancel{0} & \cancel{4} \\ -3 & 8 & 5 \\ \hline 2 & 1 & 9 \end{array}$$

5.
$$\begin{array}{rrr} & 9 & \\ 4 & \cancel{10} & 15 \\ \cancel{5} & \cancel{0} & \cancel{5} \\ -4 & 6 & 7 \\ \hline & 3 & 8 \end{array}$$

Answer to Think About It!

Let's try this problem:

$$\begin{array}{r} 800 \\ -653 \\ \hline \end{array}$$

$$\begin{array}{r} 9 \\ 7\,\cancel{10}\,10 \\ \cancel{8}\,\cancel{0}\,\cancel{0} \\ -6\;5\;3 \\ \hline 1\;4\;7 \end{array}$$

The trick is that the zero in the ones place is regrouped to become 10.

Answers to Let's Try It!

Set #3 page 49

1. 7 + 8 = 15
   8 + 7 = 15
   15 − 7 = 8
   15 − 8 = 7
2. 9 + 4 = 13
   4 + 9 = 13
   13 − 9 = 4
   13 − 4 = 9
3. 3 + 7 = 10
   7 + 3 = 10
   10 − 3 = 7
   10 − 7 = 3
4. 5 + 4 = 9
   4 + 5 = 9
   9 − 5 = 4
   9 − 4 = 5
5. 11 + 9 = 20
   9 + 11 = 20
   20 − 11 = 9
   20 − 9 = 11

Answer to Think About It!

A fact family that contains a double is smaller than a regular fact family.

Look at the example:

4 + 4 = 8
8 − 4 = 4

The fours are the same, so you have only one half as many members in this fact family. Any time you have a double in an addition/subtraction fact family, there are only two members.

CHAPTER 5

Answers to Let's Try It!

Set #1 page 57

1. 5 × 3 = 3 × 5 C
2. 2 × (6 × 1) = (2 × 6) × 1 A
3. 98 × 0 = 0 Z
4. 23 × 1 = 23 I
5. 3 × 7 = 7 × 3 C

Answer to Think About It!

Yes, it is possible. Answers will vary to show this. One possible example would be:

$$9 \times 1 = 1 \times 9$$

Answers to Let's Try It!

Set #2 page 60

1. $3 \times 5 = 15$
   $5 \times 3 = 15$
   $15 \div 5 = 3$
   $15 \div 3 = 5$

2. $8 \times 2 = 16$
   $2 \times 8 = 16$
   $16 \div 8 = 2$
   $16 \div 2 = 8$

3. $6 \times 5 = 30$
   $5 \times 6 = 30$
   $30 \div 5 = 6$
   $30 \div 6 = 5$

4. $3 \times 7 = 21$
   $7 \times 3 = 21$
   $21 \div 7 = 3$
   $21 \div 3 = 7$

5. $4 \times 6 = 24$
   $6 \times 4 = 24$
   $24 \div 6 = 4$
   $24 \div 4 = 6$

Answer to Think About It!

Your answers will vary, but they are correct if they work for both multiplication and division. A possible example:

7, 4, 28
$7 \times 4 = 28$
$4 \times 7 = 28$
$28 \div 7 = 4$
$28 \div 4 = 7$

Answers to Let's Try It!

Set #3 page 63

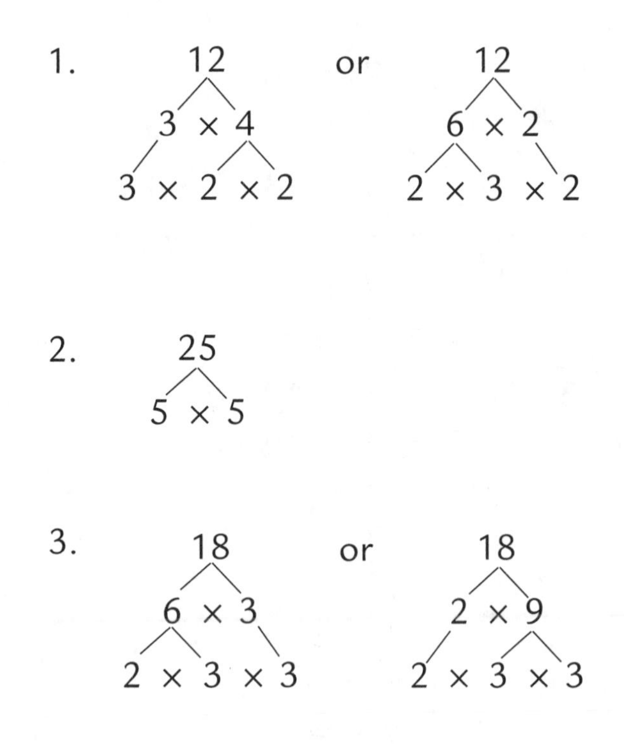

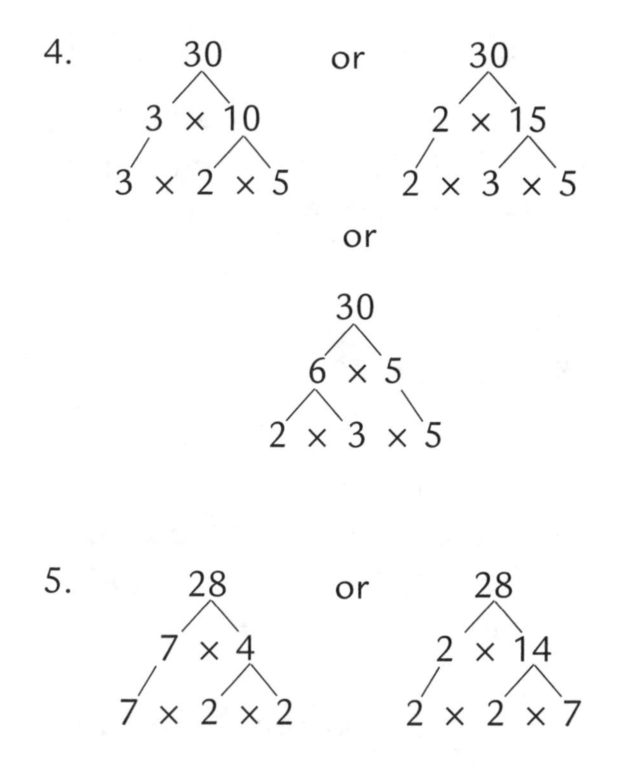

Answer to Think About It!

The tree does not always get larger. Look at this example of the tree for 48.

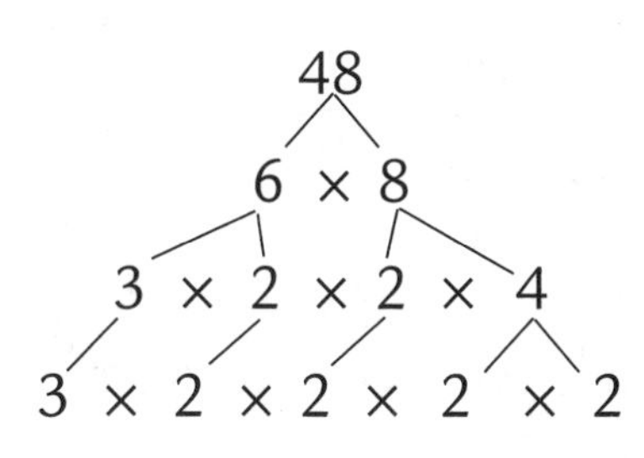

Now look at the tree for 50.

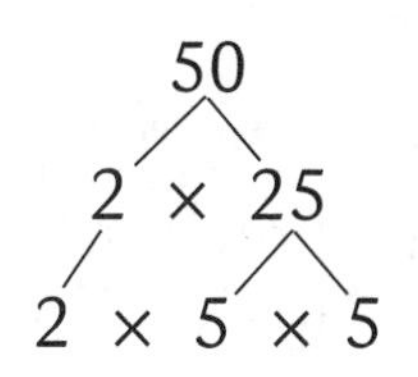

CHAPTER 6

Answers to Let's Try It!

Set #1 page 74

1. 18¢ Any three of the following:
   a. one dime, one nickel, three pennies
   b. three nickels, three pennies
   c. two nickels, eight pennies
   d. one nickel, 13 pennies
   e. 18 pennies
2. 25¢ Any three of the following:
   a. one quarter
   b. two dimes, one nickel
   c. one dime, three nickels
   d. five nickels
   e. four nickels, five pennies
   f. three nickels, ten pennies
   g. two nickels, 15 pennies
   h. one nickel, 20 pennies
   i. 25 pennies
3. 32¢ Any three of the following:
   a. one quarter, one nickel, two pennies
   b. three dimes, two pennies
   c. two dimes, two nickels, two pennies

- d. one dime, four nickels, two pennies
- e. six nickels, two pennies
- f. five nickels, seven pennies
- g. four nickels, 12 pennies
- h. three nickels, 17 pennies
- i. two nickels, 22 pennies
- j. one nickel, 27 pennies
- k. 32 pennies

4. 21¢ Any three of the following:
    - a. two dimes, one penny
    - b. one dime, two nickels, one penny
    - c. one dime, one nickel, six pennies
    - d. one dime, 11 pennies
    - e. four nickels, one penny
    - f. three nickels, six pennies
    - g. two nickels, 11 pennies
    - h. one nickel, 16 pennies
    - i. 21 pennies

5. 14¢ Any three of the following:
    - a. one dime, four pennies
    - b. two nickels, four pennies
    - c. one nickel, nine pennies
    - d. 14 pennies

Answer to Think About It!

Each equivalent set of coins for 21¢, 26¢, 33¢, or 38¢ must have at least one penny.

Answers to Let's Try It!

Set #2 page 76

1. box #1 17¢ = 17¢ box #2
2. box #1 19¢ < 25¢ box #2
3. box #1 50¢ = 50¢ box #2
4. box #1 14¢ < 15¢ box #2
5. box #1 50¢ > 46¢ box #2

Answer to Think About It!

No, because you can have fewer coins that have a greater value. For example, two quarters have a value of 50¢, but four dimes have a value of only 40¢.

Answers to Let's Try It!

Set #3 page 80

1. Julia got two pennies, one nickel, and two quarters to make 57¢.
2. Liam got one penny, two dimes, and one dollar to make $1.21.
3. Sara got two pennies, one nickel, one dime, one quarter, and one dollar to make $1.42.

4. Dana got one quarter, and three dollars to make $3.25.
5. Abby got two dimes, one quarter, and three dollars to make $3.45.

Answer to Think About It!

It takes only three coins to make 45¢. You need one quarter and two dimes.

Answers to Let's Try It!

Set #4 page 81

1. 77¢
2. 39¢
3. 66¢
4. $1.00
5. 77¢

Answer to Think About It!

It is possible to have the same answer if you have two equivalent sets of coins.

CHAPTER 7

Answers to Let's Try It!

Set #1 page 88

1. Yards
2. Inches
3. Feet
4. Miles
5. Inches

Answer to Think About It!

Inches are not really very small, so short things must be measured in parts or fractions of inches.

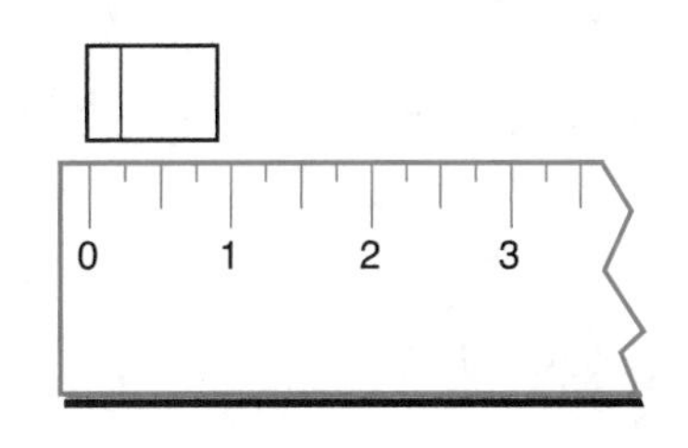

Answers to Let's Try It!

Set #2 page 93

1. millimeters
2. meters
3. centimeters
4. centimeters
5. kilometers

Answer to Think About It!

Your answer should contain this information: You use the unit that is closest in size to the object you are measuring but smaller than the object.

Answers to Let's Try It!

Set #3 page 95

1. liter
2. gram
3. milliliter
4. kilogram
5. gram

Answer to Think About It!

It does not make sense because cereal is very light. It probably weighs 35 grams.

CHAPTER 8

Answers to Let's Try It!

Set #1 page 103

1. Parallel lines
2. Intersecting lines
3. Point
4. Line segment
5. Perpendicular lines

Answer to Think About It!

No, all lines that cross are intersecting. Perpendicular lines intersect in a special way to form right angles.

Answers to Let's Try It!

Set #2 page 107

1. Parallelogram
2. Trapezoid
3. Pentagon
4. Rhombus
5. Octagon

Answer to Think About It!

Yes, because it is a quadrilateral with equal opposite sides.

Answers to Let's Try It!

Set #3 page 111

1. Cone
2. Rectangular prism
3. Sphere
4. Cylinder
5. Rectangular prism

Answer to Think About It!

A cube and a rectangular prism have the same number of faces because a cube is a special rectangular prism with square-shaped faces.

Answers to Let's Try It!

Set #4 page 114

1. One
2. One
3. Two
4. Two
5. One

Answer to Think About It!

A circle has an infinite number of lines of symmetry because a circle is formed by a path of points about one center point.

Remember that no matter how close two points are, another point can always be placed between them. So when two lines of symmetry are drawn on a circle, another one can always be drawn between them.

Answers to Let's Try It!

Set #5 page 118

1. Congruent and similar
2. Neither congruent nor similar
3. Similar
4. Congruent and similar
5. Neither congruent nor similar.

Answer to Think About It!

All circles are similar because a path of points that are all the same distance from the center point forms them.

CHAPTER 9

Answer to Let's Try It!

Set #1 page 123
(Answer shown below)

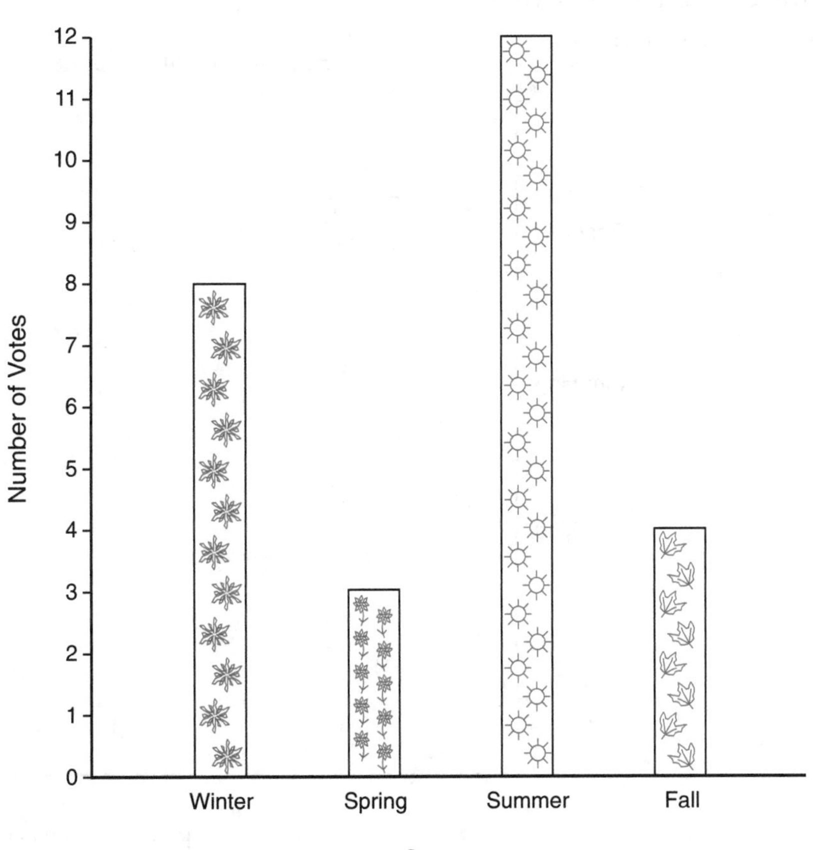

Answer to Think About It!

Add Winter with 8 votes and Fall with 4 votes to get a sum of 12 votes, which is the same or equal to the number of votes for Summer.

Answers to Let's Try It!

Set #2 page 125
(Answer shown below)

Answer to Think About It!

The data will be different because students who live in the Midwest will not be able to surf, so none of them will choose surfing.

Answers to Let's Try It!

Set #3 page 127
(Answer shown on facing page)

Answer to Think About It!

Answers may vary, but they should include something about the time of year. Many students are not in school for much of the month of June so collection of newspapers will be slower. The collection for July will be even less than for June, if the trend continues.

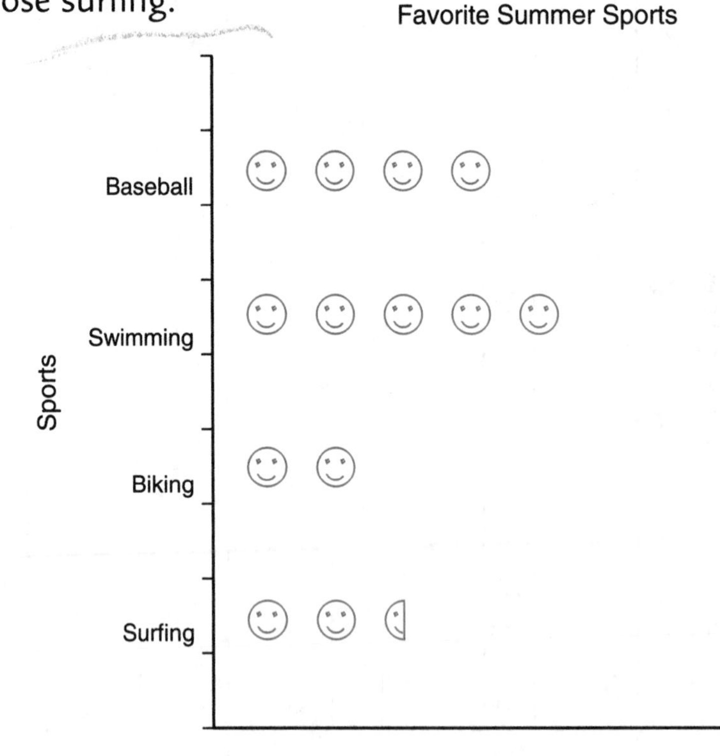

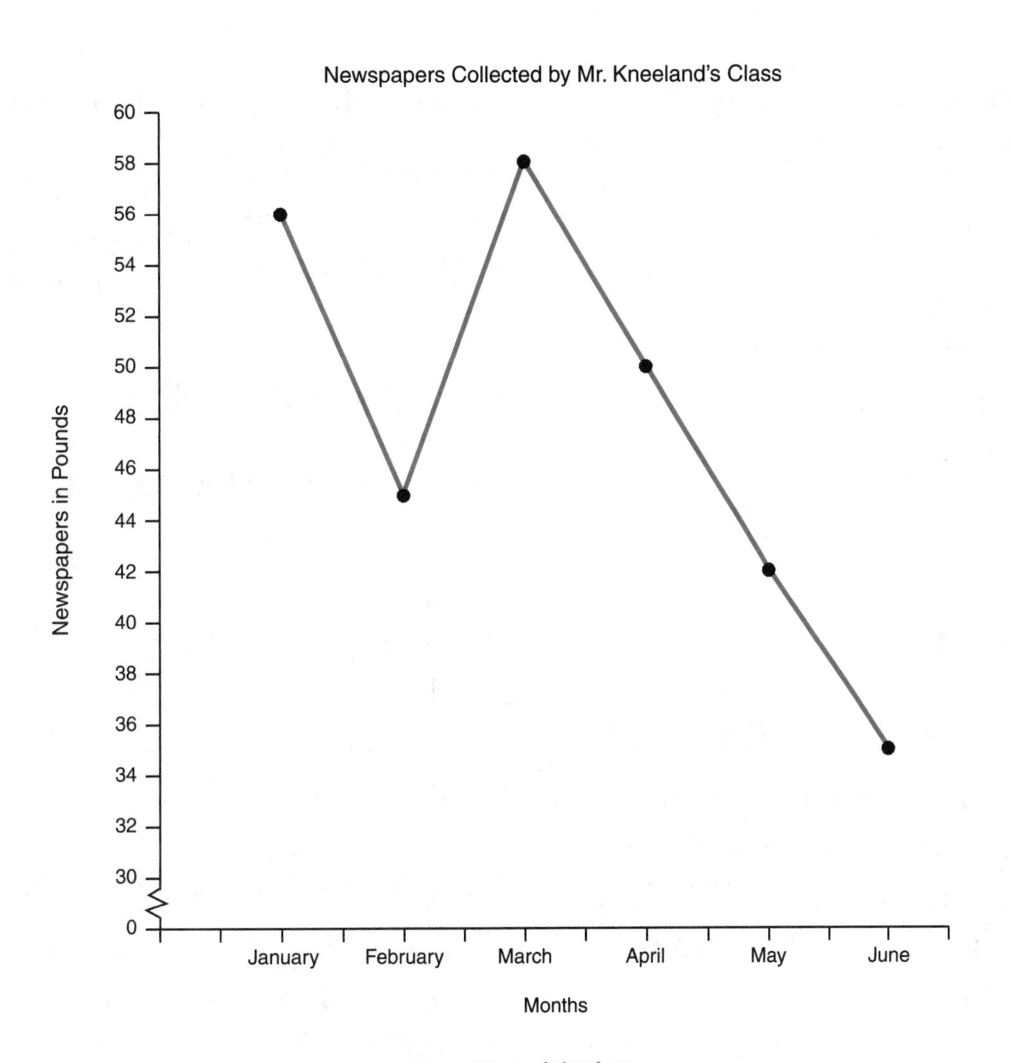

CHAPTER 10

Answers to Let's Try It!

Set #1 page 136

1. $\frac{2}{6}$ or $\frac{4}{12}$
2. $\frac{2}{10}$
3. $\frac{1}{4}$ or $\frac{2}{8}$
4. $\frac{6}{8}$ or $\frac{9}{12}$
5. $\frac{1}{4}$ or $\frac{3}{12}$

Answer to Think About It!

The pizzas are the same size, so the meatball pizza has bigger pieces because there are only six pieces. The pepperoni pizza is cut into eight pieces, so each piece is smaller.

Answers to Let's Try It!

Set #2 page 139

1. $\frac{2}{9}, \frac{3}{9}, \frac{5}{9}, \frac{7}{9}$
2. $\frac{1}{12}, \frac{2}{12}, \frac{3}{12}, \frac{5}{12}, \frac{8}{12}$
3. $\frac{3}{9}, \frac{3}{8}, \frac{3}{5}, \frac{3}{4}, \frac{3}{3}$
4. $\frac{6}{8} = \frac{3}{4}$
5. $\frac{1}{3} > \frac{1}{4}$
6. $\frac{2}{5} < \frac{2}{3}$

Answer to Think About It!

Use the fraction bars and the ruler. Line up the ruler so that it runs up and down. Move it from left to right. The first fraction you pass is the least.

Answers to Let's Try It!

Set #3 page 141

1. $\frac{3}{8}$
2. $\frac{7}{10}$
3. $\frac{6}{9}$
4. $\frac{4}{12}$
5. $\frac{5}{6}$

Answer to Think About It!

The sum is $\frac{8}{8}$, which is the same as **one whole**. It looks like this:

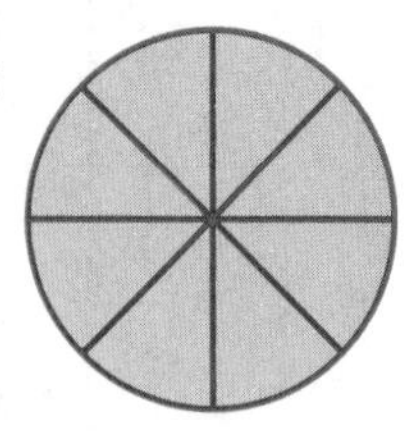

CHAPTER 11

Answers to Let's Try It!

Set #1 page 147

1. Impossible
2. Certain
3. Likely
4. Unlikely
5. Likely

Answer to Think About It!

Answers will vary. Check to see if an event is truly certain or impossible. This may be the hardest to write. As a matter of fact, that's certain!

Answers to Let's Try It!

Set #2 page 150

1. $\frac{3}{5}$ or three chances of five
2. $\frac{1}{6}$ or one chance in six
3. $\frac{3}{6}$ or three chances in six
4. $\frac{3}{8}$ or three chances in eight
5. $\frac{2}{6}$ or two chances in six

Answer to Think About It!

The probability of spinning a blue is 0 because there is no blue on the spinner.

CHAPTER 12

Answers to Let's Try It!

Set #1 page158

1. Yes, all four pictures will fit on Alvaro's display.

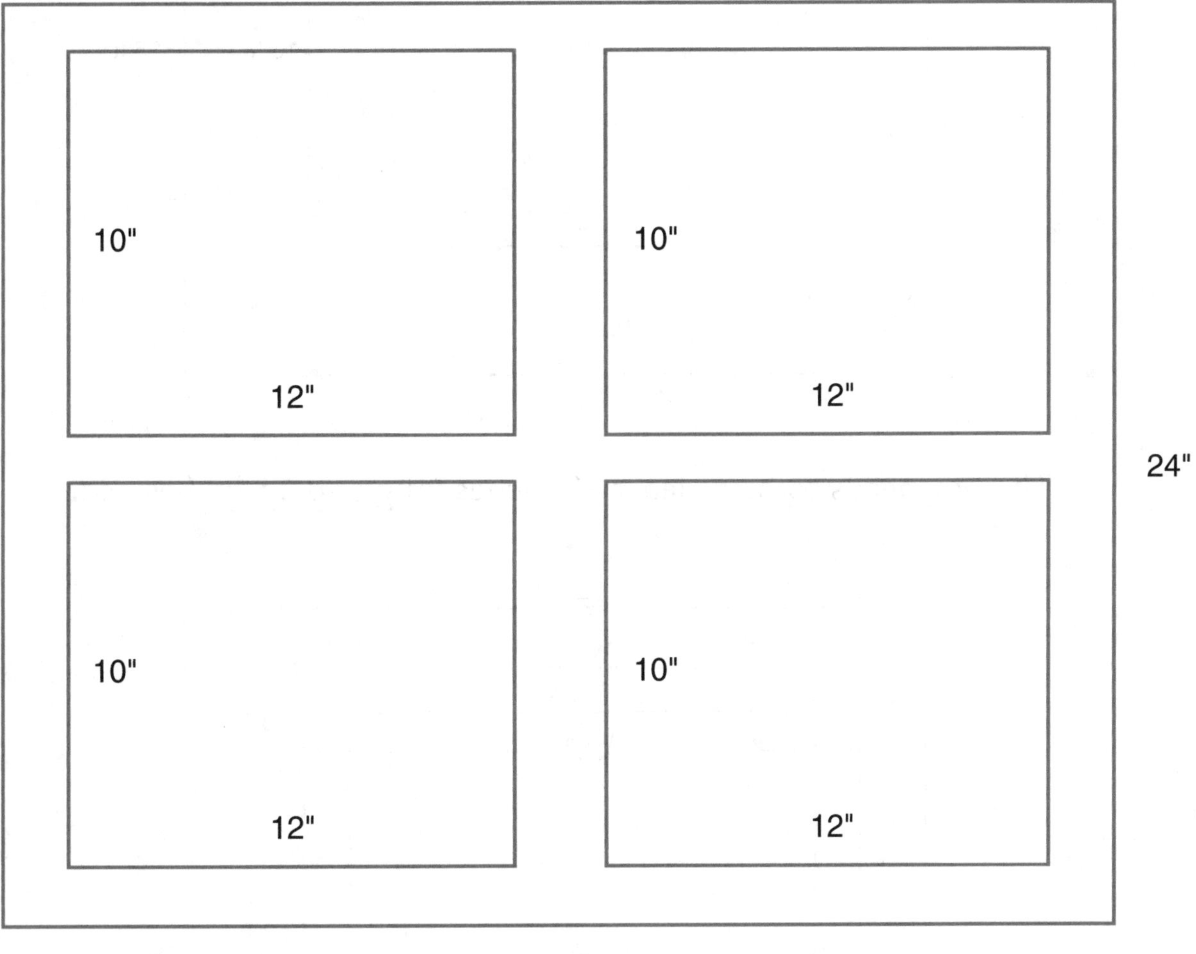

2. The fence will be 30 feet long and 25 feet wide.

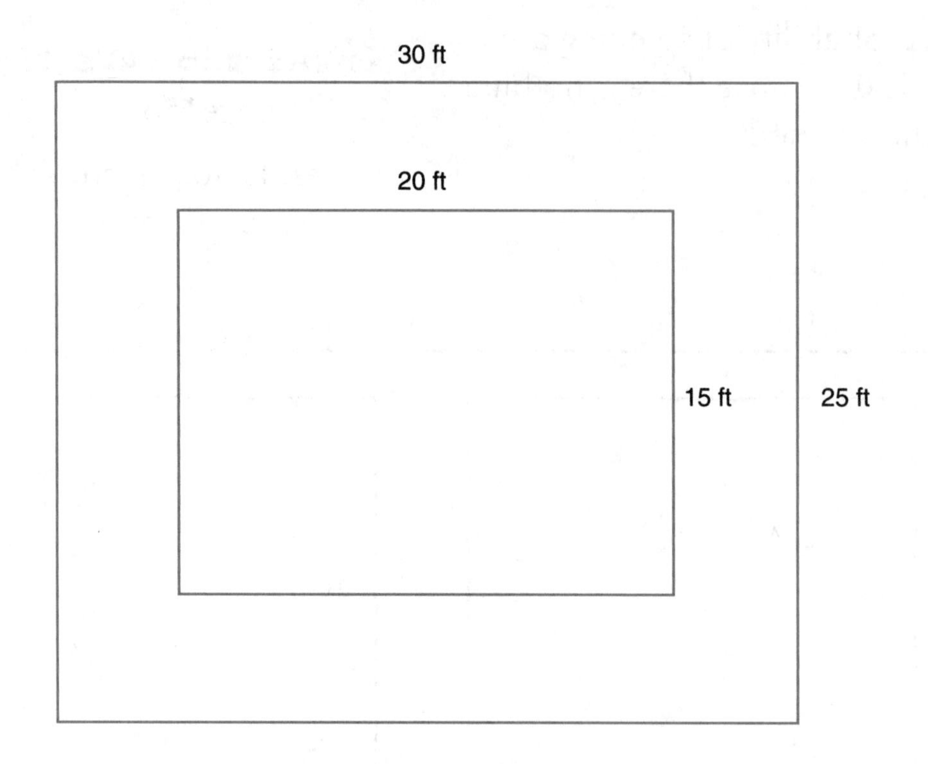

3. The box needs to be at least 18 inches long and 15 inches wide.

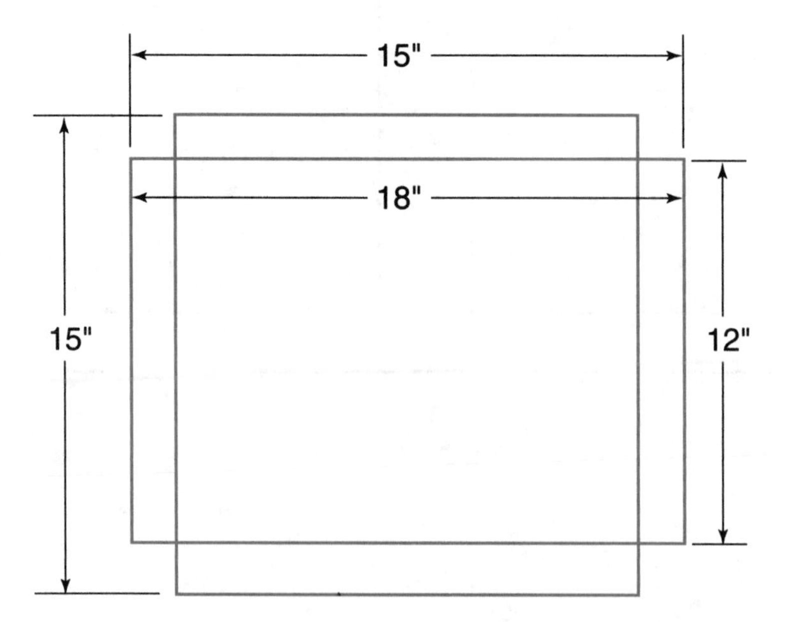

Answers to Let's Try It!

Set #2 page 160

1.

shirt	shorts	cap
pink	brown	yellow
pink	brown	green
pink	white	yellow
pink	white	green
blue	brown	yellow
blue	brown	green
blue	white	yellow
blue	white	green

Rebecca will have eight different outfits at camp.

2.

bread	filling	spread
roll	ham	mustard
roll	ham	mayonnaise
roll	salami	mustard
roll	salami	mayonnaise
rye	ham	mustard
rye	ham	mayonnaise
rye	salami	mustard
rye	salami	mayonnaise

Kathryn can make eight different kinds of sandwiches.

3.

ice cream	topping	cone
chocolate	cherry	sugar
chocolate	cherry	waffle
chocolate	sprinkles	sugar
chocolate	sprinkles	waffle
vanilla	cherry	sugar
vanilla	cherry	waffle
vanilla	sprinkles	sugar
vanilla	sprinkles	waffle

Donna can make eight different ice cream cones.

Answers to Let's Try It!

Set #3 page 162

1. At the end of each week, Nora is making 50¢ more than the week before. At this rate, Nora makes $3.50 at the end of the fifth week, so at the end of the sixth week she will earn $4.00.
2. The pattern here is to add seven days or one week to each concert date. At this rate, the fourth concert will be given on July 23 and then the fifth concert will be held on July 30.

3. Brenda's arrangement adds three pinecones to each row. The fourth row will have 14 pinecones, and the fifth row will have 17 pinecones.

Answers to Let's Try It!

Set #4 page 164

1. To solve Millie's pizza party problem you can cut some papers for each of the tables. Put a five on some of them and a six on others. Add combinations of five and six until you get three tables of six and one table of five. You will have a total of 23 people sitting at tables with all tables full.

2. To solve Livvie's t-shirt problem draw a grid that has three rows and two columns. Cut six small pieces of paper that fit into the boxes formed by the grid. As you read the clues, place the color words on the grid to solve the problem. Your solution will look like the one below.

Green	Blue
Pink	Yellow
White	Red

3. To solve Allan's backpack problem, cut some pieces of paper and label some of them with the number 4 and some with the number 5. Try different combinations of four or five to get to 17. If you do it right, you will have three pieces of paper with the number 4 and one paper with the number 5, so Allan needs three backpacks that hold four lunches and one backpack that holds five lunches.

Answers to Let's Try It!

Set #5 page 166

1. Set up a table of the hiking trips and place the trails in order from the shortest to the longest.

   This table shows the trails from the shortest to the longest, so this is the order that Anne should hike them.

Trail	**Length in Miles**
Elegant Fir	6
Hound's Tooth	8
Long Beach	9
Quiet Elm	10

2. Set up a table showing days that Jack and Duncan go to batting practice.

Jack	1	2	③	4	5	⑥	7	8	⑨	10	11	⑫
Duncan	1	2	3	④	5	6	7	⑧	9	10	11	⑫

   The day that each boy goes to the batting cage is circled. On the 12th day both boys will go to the batting cage.

3. Set up a table for the turtle's and the frog's trip.

Turtle	1 inches	2 inches	4 inches	8 inches	16 inches	32 inches
Frog	1 inches	7 inches	13 inches	19 inches	25 inches	31 inches

Once you put the numbers in the table, you can see that the turtle makes it to the pond. The frog is 1 inch away. The turtle wins!

Answers to Let's Try It!

Set #6 pages 167–168

1. If there were more than five people at the party, let's begin by guessing six people.

   30 ÷ 6 = 5

   That won't work because there is nothing left. Let's guess seven people at the party.

   30 ÷ 7 = 4 R 2

   That won't work because there is not enough left over as the remainder. Let's guess eight people at the party.

   30 ÷ 8 = 3 R 6

   That works! So the answer is eight people came to Clara's party.

2. We can start with 18 red marbles.

   24 − 18 = 6

   Since 18 is not twice 6 those two answers won't work. Let's try 17 red marbles.

   24 − 17 = 7

   That won't work either. 17 is not twice 7, but we are closer. Let's try 16 red marbles.

   24 − 16 = 8

   That's it! So the answer is 16 red marbles and eight black marbles in the bag.

3. To answer this question, we need two numbers that are 1 away from each other and that, when multiplied, will have a product that is 5 less than 47. Let's try $4 \times 5 = 20$. That won't work. Let's try $5 \times 6 = 30$. That won't work either, but we are closer. Let's try $6 \times 7 = 42$. If we add $42 + 5$ we get 47! That's the answer! There are seven shelves with six books on each shelf and five books left over.

Answers to Let's Try It!

Set #7 page 169

1. If the practice ended at 12:30 P.M, go back 45 minutes in time to get to 11:45 A.M.
2. To find out how much Ona spent on her guitar case we start by adding the money she spent on lunch to the money she had left.

   $10 + $17 = $27

   Now subtract that from what she started with.

   $100 - $27 = $73

   Ona spent $73 on her guitar case.
3. We can find out how much the meat in Arny's sandwich weighed by first adding the other parts.

   2 oz + 4 oz = 6 oz

   Now subtract that from the total weight.

   11 oz − 6 oz = 5 oz

   Arny used 5 ounces of salami.

Answers to Let's Try It!

Set #8 page 170

1. Four girls must be arranged. We can make it simpler by trying only three of them. It will look like this using just the first initial.

   JEM EMJ MEJ
   JME EJM MJE

   Now we can add the last initial. We will look at how many can be made using J first.

   JEMS JESM JMES
   JMSE JSEM JSME

   There are six different arrangements with Jamie in first place. To find the total number of arrangements multiply 6 times 4 to get the total of 24 different arrangements.

2. The Westerhoff family's trip to the show is complicated by ticket prices that are not on the dollar. We will simplify the ticket prices so that the adults' tickets cost $9 and the children's tickets cost $5.

   2 × $9 = $18
   and
   3 × $5 = $15

   $18 + $15 = $33

   The actual price should be close to this number.

   So

   2 × $8.78 = $17.56
   and
   3 × $5.49 = $16.47

   $17.56 + $16.47 = $34.03

   This answer is very close to our simpler problem, so it makes good sense.

3. The number of members in Mrs. Deger's art club and the number of drawings are not easy to use, so we will make them simpler. We will change the number of drawings to 50 and the number of art club members to 50 as well. To find the total number of drawings made, we will multiply.

   50 × 50 = 2,500

   So the actual answer should be close to 2500 drawings. Let's try the actual numbers.

   55 × 47 = 2,585 drawings

   This answer makes sense because it is close to our simpler problem.

Index